TREES, WOODS AN

Trees, Woods and Forests

A Social and Cultural History

Charles Watkins

REAKTION BOOKS

Published by Reaktion Books Ltd
Unit 32, Waterside
44–48 Wharf Road
London N1 7UX, UK

www.reaktionbooks.co.uk

First published 2014, reprinted 2016
Paperback edition first published 2016

Copyright © Charles Watkins 2014

Printed and bound in Great Britain by Bell & Bain, Glasgow

A catalogue record for this book is available from the British Library

ISBN (hb) 978 1 78023 373 4
ISBN (pb) 978 1 78023 664 3

CONTENTS

Introduction *7*

ONE Ancient Practices *17*

TWO Forests and Spectacle *36*

THREE Tree Movements *65*

FOUR Tree Aesthetics *91*

FIVE Pollards *119*

SIX Sherwood Forest *140*

SEVEN Estate Forestry *175*

EIGHT Scientific Forestry *204*

NINE Recreation and Conservation *224*

TEN Ligurian Semi-natural Woodland *247*

Afterword *269*

REFERENCES *277*
SELECT BIBLIOGRAPHY *298*
ACKNOWLEDGEMENTS *301*
PHOTO ACKNOWLEDGEMENTS *303*
INDEX *305*

But when I walk past our village woodlands which I've saved from the axe or hear the rustle of my own saplings, planted with my own hands, I feel I too have some control over the climate and that if man is happy a thousand years from now I'll have done a bit towards it myself.

Dr Astrov in Anton Chekhov, *Uncle Vanya*,
Act I (1897), trans. Ronald Hingley

The ash is quite bare. Oddly enough the last leaves seem to linger on the top rather than lower down, as one might have expected. It's really shocking, I am becoming a nature-lover and observer – fatal! – The intellect fades in proportion.

Lytton Strachey to Roger Senhouse, 12 November 1928

What kinds of times are they, when
A talk about trees is almost a crime
Because it implies silence about so many horrors?

Bertolt Brecht, 'To Those Born Later', *c.* 1938

# Introduction

O UR understanding of the history of trees and woodlands has been transformed in recent years. Established ideas, such as the spread of continuous dense forests across the whole of Europe after the Ice Age, have been questioned, if not overturned, by archaeological and historical research. While there is continued clearance of tropical forests, concern over woodland loss in Europe, where the area of land covered by trees has increased substantially in the last century, is less well founded. Recent research shows that the interactions between humans and trees and woods have varied dramatically through time and from place to place. Over history the clearance of woodland has often been celebrated as a sign of increasing population; an indication of improved agricultural production; a surrogate for civilization. But woodland is not a simple category from which a settled landscape has been wrought. Rather it is a complex category which has varied dramatically in the density, age, species and form of its trees and shrubs. The utility of woodland and the cultural values ascribed to it are also diverse, whether it is tilled or grazed by domestic stock: it may be a provider of status or symbolic power, a site of traditional management or scientific experimentation.

A key issue is that the same trees and woods are perceived very differently by different people at the same time, and by groups of people through time. John Ruskin argued that a love of trees could be a moral test failed by those who disliked them, but most people like trees in some circumstances and dislike them in others. And the reasons for like or dislike can sometimes be almost unfathomable due to personal recollections and associations with trees. I vividly remember being shown in 1978, from a library window, an attractive nineteenth-century landscape park in the English Midlands scattered into the distance with a diverse range of plantations and groves. While I was appreciating the well-tended young plantations of pine and the old oaks, the owner pointed out with relish that the plantation to the left was under Schedule 'D' for income tax purposes and the clumps of oaks to the right were Schedule 'B', while the large wood in the middle distance was a mixture of both. The view was revealed as a landscape of tax avoidance. The principal reason the woods were so carefully tended was that the movement of woods between schedules allowed heavy expenditure incurred during replanting to be offset against income from other sources. The beauty and productivity of the woodland was a consequence of the then tax regime, which had been introduced in 1915 to encourage woodland management. This meaning was invisible to anyone uninitiated into the intricacies of tax accounting.

The aim of this book is to examine how different understandings and values ascribed to trees, woods and forests help to provide rich layers of association and meaning to them. The pattern of trees in the landscape is the result of centuries of unremitting hard work by our ancestors as they struggled to survive in harsh conditions. The work of management and control of trees has involved careful shepherding of flocks of sheep and goats, manipulation of fire and skilled harvesting of leaves, fruits, seeds and branches. Woodland can be benign and provide shelter, warmth, sustenance, fuel and fodder. But it may also be dark, disorientating and threatening and associated with debauchery and death. The landscape of woods and forests is imprinted with ancient patterns of power and desire.

This book explores our understanding and knowledge of trees, woods and forests by examining a series of episodes and moments in woodland history. It draws on a wide array of sources and approaches. The history of trees and woods is intrinsically interdisciplinary; research is carried out by archaeologists, botanists, ecologists, foresters, geographers and historians. Much of the most interesting and fruitful research is being undertaken at the edge of disciplines and uses approaches from cultural history, historical ecology, local history, natural history, historical geography and field ecology. Very different forms of evidence are combined to develop an understanding of the history and development of a particular piece of vegetation. This evidence may include surveys of present-day plants and animals; historical maps and documents; field archaeology; dendrochronology; oral history; land management practices; literary descriptions; genetic analysis; pollen and soil analysis; aerial photographs; and drawings and paintings.

The shape and form of a tree is itself one of the best sources of evidence of its history. It can tell you whether a tree has been lopped or coppiced in the past and whether it has been affected by diseases, grazing or browsing. Trees are usually defined as having some sort of perennial, woody growth which often, but not always, has a main stem or trunk with woody branches. They are enormously varied: some species, such as the Californian coast redwood (*Sequoia sempervirens*), can grow over 100 metres high and form dense stands of ancient trees. By contrast, the alpine woods of the Hayachine mountain in northern Honshu consist of 1- to 2-metre-high species such as *Pinus pumila* and *Tsuga diversifolia* hugging the ground in dense low mats of interlocking vegetation where it is difficult to distinguish individual trees (illus. 66). Trees are usually rooted and static, but some trees growing along rivers and in river beds are moved by floods. Trees can extend their range rapidly through windblown seeds, or by fruits and nuts carried by animals and rivers; they may also spread several metres a year through the stealthy creeping growth of root suckers. Many trees are ancient beings hundreds of years old, yet the oldest trees are not always the largest: within a wood, small, stunted trees which grow very slowly can be the same age as neighbouring large trees of the same species which reach to the canopy (illus. 65).

The ideas associated with the terms 'woods' and 'forests' vary through time and from country to country. Recent research in historical ecology is increasingly showing that the boundaries between woodland, wood pasture, pasture with trees and arable land are difficult to define with precision. In general terms, distinct boundaries between dense woodland and more open woodland have become more pronounced in many places over the last 300 years. These changes are related to the more intensive management of land and the greater control over grazing, and are very closely tied to changes in the pattern of land ownership and control. But there remain many areas of semi-natural vegetation where it is difficult to map distinct boundaries between woodland and non-woodland.

These definitional ambiguities emphasize the great care needed in interpreting the history of woods and forests. The changing uses of the terms can obscure; the changing meanings themselves can, however, be used to uncover changing attitudes and practices. In England, for example, the medieval Royal Forests were areas where the monarchy retained special hunting rights. All included land which was deemed suitable for hunting, but some areas, such as Exmoor, had far fewer trees than others such as the Forest of Dean. Most forests were made up of tracts of land which could contain villages, heaths, arable land, pasture and woodland. There was no direct connection between the idea of forest and the concept of woodland: medieval forests were administrative units more akin to a modern national park than extensive areas of planted trees. With the decline in Crown interest, especially from the eighteenth century onwards, the term 'forest' became increasingly associated with those wooded areas, such as the New Forest and the Forest of Dean, which survived as Royal Forests. It was not, however, until the establishment of professional forestry in the nineteenth century, and the intermixture of traditional estate woodland management with ideas of scientific forest management introduced from the Continent, that the terms 'forestry' and 'forester' began to be the normal terms used to describe woodland management and managers. It was the state Forestry Commission, established in 1919, which introduced the word 'forest' to describe its administrative units and then went on to use the term 'National Forest Park' for recreational areas. Their use of 'forest' was ambiguous: it linked the coniferous afforestation of massive areas of both upland and lowland with ideas of ancient deciduous woods associated with the remnant Royal Forests.

Trees and woods often outlive humans and provide a semblance of order, continuity and security. The sense of regret brought about by the sudden loss of familiar trees can be acute. In the late 1960s and early '70s there was growing concern in lowland England about the loss of broadleaved woodland and hedgerow trees through agricultural expansion and the enlargement of fields. This was exacerbated in the early 1970s by the devastating effect of Dutch elm disease, which stripped most of England bare of one of the most important trees of hedgerow and small woodland.[1] The agricultural landscapes of vast tracts of the country were rapidly denuded and skeins of intertwined dead English elms (*Ulmus procera*, which had been introduced by the Romans) and wych elms (*U. glabra*) remained for several years as memorials of the implications of the global spread of tree diseases.

This concern over woodland loss and decline was more than matched by concern over the effects of large-scale coniferous afforestation in the uplands of Britain. This showed itself in the Lake District in the late

1930s, where there were intense debates over the darkening spread of introduced conifers that was so clearly visible across the fells. There was less of an outcry over similar though largely invisible afforestation schemes over extensive areas of flat, lowland heathlands such as Sherwood Forest and the East Anglian Breckland. The change in the appearance of landscape was mitigated by changes to the planting schemes of the Forestry Commission in the 1950s and '60s, but the rapid increase in public and private coniferous afforestation in the 1970s and early '80s heightened the level of concern. Some of the keenest critics of afforestation were organizations, such as the Ramblers' Association, who saw the new plantations as restricting public access to the open heaths and moors. The argument was fiercest in Scotland and Wales, and the planting of Sitka spruce over areas of international significance for bird conservation, such as the Flow Country of Caithness and Sutherland, brought matters to a head in the 1980s and led to afforestation restrictions.[2]

By the 1980s public concern, initially in Central Europe and eastern North America, became acute concerning the potential effects of 'acid rain' on trees and woods. Scientists identified a forest malaise which they termed Waldsterben, a very worrying general forest decline brought about by air pollution. Many popular articles were written and illustrated by images of dying and dead trees. But within less than ten years careful research showed that there was no such general decline. Although more chimera than fact, and now a largely forgotten moment in forest history, the acid rain 'crisis' was important in sustaining the level of public concern over the fragility of trees and woodlands. In England, this was further reinforced by the Great Storm of 15–16 October 1987, the worst within living memory, which had a particularly devastating effect on trees in southeast England. These threats to trees and woods loomed large in the public imagination in Britain, and were part of a worldwide concern over loss of tropical rainforests through logging and conversion to grazing land.[3]

Another key debate over woodland management concerned the rapid rise in importance given to the nature conservation and cultural value of semi-natural habitats. The insights provided by historical ecology showed that many woods were of ancient origin and that traditional forms of management such as coppicing, which had largely stopped in the interwar years, in many instances had beneficial effects on nature conservation. This nuanced approach to woodland management had important implications for policy, especially on the pressure to stop the replanting of old woodland sites with conifers. The category 'ancient woodland' was formulated to engage the public and help preserve and then conserve areas that had been woodland for hundreds of years. There is now a general consensus that ancient woodlands should be protected, although vigilance is still required to safeguard them from new roads, railways and houses.[4]

If there was any doubt of the very high regard with which trees, woods and forests are held by many people and the strong resonance they have with ideas of local and national identity, this was ousted by the vociferous campaign in February 2011 to stop plans by the government to sell woodland owned and managed by the Forestry Commission. The campaign raised considerable concerns about the provision of public access to woodland and the benefits that accrue to society from different types of woodland management. A group called Save England's Forests rapidly gained the support of the Archbishop of Canterbury, Annie Lennox, Dame Judi Dench and other celebrities. The group wrote a letter to the *Sunday Telegraph* (22 January 2011), made extensive use of Facebook and Twitter and achieved its aim with remarkable ease; within weeks the government reversed its policy of selling off public woodlands.

The high profile and success of this campaign and the evidence collected by the consequent Independent Panel on Forestry emphasizes deep public support for trees and woodland and the wide range of benefits they provide, including timber production, landscape and culture, wildlife and game conservation, public access and shelter. Woodlands are increasingly valued for their ability to reconnect children and adults with nature; they are seen as a type of therapeutic landscape. Knowledge of the precise and specific role of trees and woodland soils in providing habitats for mosses, lichens, fungi, insects, birds and mammals is improving rapidly but at the same time local traditional knowledge is being lost.[5] In the last ten years or so an additional value, that of carbon storage, has been ascribed to woodland, providing yet another reason for woodland establishment and management.

The relationship between humans and trees is rich, intricate and multilayered. This book examines the different ways in which people have used, understood and appreciated trees, woods and forests through a close examination of episodes and moments in woodland history. The emphasis is on Britain but examples are drawn from across the world. The approach is broadly chronological but chapters are structured around key themes and periods which shed light on changing understandings of trees and woods. Chapter One considers the importance of ancient tree and woodland management practices such as coppicing and the evidence for tree and woodland management from archaeology and ancient literature. It emphasizes how essential trees were for the survival of our prehistoric ancestors and the specialized knowledge and expertise that they had. Trees have been managed for thousands of years and the labour and effort of our ancestors were essential in creating the soils and landscapes which we cultivate. Chapter Two focusses on forests and spectacle and the way in which the expression of power and the desire for understanding the origin of the world are related to trees. The symbolic importance of forests for hunting and its association with military prowess and horsemanship were vital for Alexander the Great and Roman emperors such as Hadrian. Trees and woods are closely associated with classical, Christian and Norse creation myths. The Norman kings of England used forests to express their power and produce crucial income.

Humans have moved trees from place to place for centuries and have rapidly revolutionized the distribution of individual species. The speed of movement has increased dramatically over the last 400 years. Chapter Three examines the fascination and enthusiasm of European collectors from the seventeenth century onwards and how new species were collected, named, catalogued, tested and acclimatized. The great enthusiasm for novel trees led to the establishment of important tree collections, which from the early nineteenth century became termed 'arboreta'. The chapter concludes with a case study of the Japanese larch, which after its introduction into Britain rapidly became a common plantation tree, but is now threatened by an introduced disease.

Trees are not only crucial for timber, fuel, food and fodder; people have long valued them for their beauty. Many different types of trees are appreciated. Some value the tall, clean lines of large standard trees such as beeches, oaks or pines grown in plantations or limes and oaks as avenues along roads or across parks. Others enjoy ancient, crooked oaks, ashes and thorns growing in hedgerows or in remnant areas of wood pasture and commons. Chapter Four examines how trees and woods have been represented in art and how attitudes to trees were transformed by the rise of the cult of the Picturesque in the eighteenth century. Chapter Five concentrates on the ancient practice of pollarding and its importance for leaf fodder for farm animals. The last vestiges of the practice in Greece are considered, as are changing attitudes to pollarding in Britain, which were strongly influenced by questions of aesthetics and power.

The same trees and woods are perceived very differently by different people at the same time, and by groups of people through time. Chapter Six examines these different meanings and values by exploring changing interpretations at Sherwood, one of the world's most famous forests, which rose to international fame at the start of the nineteenth century just as its legal identity was lost. The changing ways in which the ancient oaks of Sherwood have been celebrated and enjoyed are identified. Chapter Seven focusses on British forestry in the nineteenth century, when most landed estates had a wide variety of trees and woods varying from hedgerow trees and small spinneys within agricultural land to larger woods managed for timber and trees and woods in parkland. The value of different types and patterns of woodland for game shooting and fox hunting are analysed, as are the ways in which trees were used in landscaping and to naturalize, disguise or emphasize property ownership.

The invention of the idea of sustainable woodland management and forestry is described in chapter Eight. This opens with a consideration of late medieval ideas of sustainable woodland management in northern France and the development of scientific forestry in Germany. The spread of these ideas into India and America in the later nineteenth century and to Britain after the First World War is examined, as are some of the landscape implications of scientific forestry and consequent large-scale afforestation of land formerly used for grazing animals. Chapter Nine examines the repercussions of the 'discovery' of Californian redwoods for the development of forest conservation and the concept of national parks and public reserves. It then considers how the idea of National Forest Parks was developed in Britain and how it was used to help normalize large-scale newly established plantations of conifers.

The final chapter considers the spread of unplanned, naturally regenerated woodland as a contrast to afforestation. Very large areas of woodland have become established in the Italian Apennines over the last century through the abandonment of former agricultural land and pastures and the consequent natural regeneration of trees and shrubs. Chapter Ten considers the reasons for this process and the effects on cultural landscapes and nature conservation. The abandonment of the complex and subtle mix of activities and landscapes that includes the grazing of summer pastures, agricultural terraces, tree management and chestnut culture has led to massive natural regeneration of trees and woods and the loss of the former cultural landscape. This new 'rewilded' landscape poses significant questions for the conservation of nature and heritage. While some people celebrate letting land go out of cultivation and the provision of apparently wild and untamed landscapes that encourage the spread of wild boar, wolves and eagles, others decry the loss of local meaning and subtlety and the coarsening of the landscape.

The close examination of particular episodes, moments and themes that forms the basis of this book should help to elucidate and deepen an appreciation and understanding of trees, woods and forests. Trees can live to a great age, and some woods appear timeless and ancient, yet they are also prone to sudden death through fire and storm. Trees can live for many hundreds of years if they are repeatedly browsed and grazed by animals. Human activity has led to the clearance and destruction of enormous areas of woodland and has also brought about the rapid spread across continents of potent diseases such as chestnut blight, elm disease and ash dieback. Humans have also created vast areas of forest plantations and have managed the form and shape of individual trees for centuries. Trees and woods are celebrated for their fixedness in the landscape and used as landmarks and sites of memory. Trees are planted and grow thousands of miles from the places where they originated, and tree seeds rapidly spread and grow over abandoned fields and villages. In this way trees, woods and forests represent change and mark episodes in history.

1 Charles Holroyd, *Eve and the Serpent,* 1899, etching.

ONE
Ancient Practices

THE complex relationships between humans and trees go back many thousands of years. There was a time when trees were of vital importance for almost all human activities: making clothes; providing food, fuel and fodder; constructing houses; making tools, weapons and wheels; providing shelter and shade. One of the most exciting archaeological finds of recent years that demonstrated the dependence of early humans on trees and shrubs was the extraordinary discovery of a frozen man at the Similaun glacier in the Tyrolean Alps on the borders of Austria and Italy in September 1991. The body has been dated to 3300 BC and the thorough examination of the body, the man's clothing and belongings and the site raises many questions as to his social status. He has been interpreted as an outlaw, a hunter or warrior, a priest, an ore prospector and, perhaps most likely, a shepherd making use of the high summer pastures in the Ötz valley. While his way of life remains contentious, the preservation of his equipment and clothing allows a precise and accurate archaeological interpretation of the many varied uses to which different tree products were put.[1]

Wood was a key component of most of the surviving clothes and artefacts owned by the ice man and he made use of six main species of tree. The lime (*Tilia* spp.) was used for the greatest variety of purposes. The inner bark of the lime, called bast, can be separated into fibres that can then be twisted to make string and rope. This bast was used as a sewing material for his shoes and for various containers. It was also used to make the string which formed the basis of his backpack and the sheath for a dagger. A very specialized use of lime was discovered in a pouch 'which held two blades and a borer made of flint, an awl made from a sheep or goat bone, pieces of true tinder fungus (*Fomes fomentarius*) and a previously unknown tool for sharpening flints' which consisted of 'a completely debarked branch of lime in which a peg of a deer antler was inserted'. This tool was just over 11 centimetres long with a diameter of 2.6 centimetres, and was used for making or sharpening flint tools.[2]

Yew was used for the ice man's two main weapons, his bow and his axe. The use of yew for the longbow is hardly surprising as this was for many centuries the favoured wood for this purpose: all prehistoric bows are of yew and as recently as the sixteenth century large quantities of yew were imported from the Tyrol for the English Army. The ice man's bow was in the process of being made and has many whittling marks, showing that it was not ready for use. More unusual was the use of yew for the handle of his axe, since most surviving axe handles of the same period are made of ash, oak, beech or pine. The ice man's axe handle was made from a 'longish piece of trunk from which a strong branch stuck out almost at right angles'. The trunk of yew was trimmed to make the haft of the axe and the branch was made into the shaft that held the copper blade. This method of axe making has a long tradition stretching from the Neolithic to the Iron Age. The only use that the ice man had for ash was as the handle of a dagger. This is surprising, as ash wood had many purposes in prehistoric times, since it is pliable and relatively easy to work into tool handles.[3]

Hazel stems are very strong and pliable and were used by the ice man to construct frameworks for his backpack and quiver. The frame of his haversack was 'constructed from a thick branch of hazel (*Corylus avellana*) bent into a U-shape' with 'two coarsely-worked laths of larch (*Larix decidua*)'. The hazel spar had been stripped of its bark and side branches. It was notched, and these notches probably helped to hold the larch laths in place. Later a third larch lath was found nearby. The goatskin quiver, which was given structure by hazel stems, held fourteen arrows made from straight, thin yet dense and hard stems of the wayfaring tree *Viburnum lantana*. Most of the arrows were unfinished, like the bow, and a broken arrow had been repaired, with the front end replaced with a stem of dogwood (*Cornus*). The ice man carried two containers made of birch bark, which is fairly easy to detach from young trees and remains flexible and strong. These characteristics made it 'an ideal material for the manufacture of containers and cases' while the sap of the birch was used to make the glue which helped fix the axe blade in place. One of the containers was used to carry embers; leaves of the Norway maple (*Acer platanoides*) were used as insulating material

2 Reconstruction of a birch-bark vessel found with 'Ötzi', the Tyrolean ice man (3300 BC).

between the hot embers and the wall of the birch bark container. The embers consisted of a mixture of charcoal including spruce or larch, pine, green alder (*Alnus viridis*), elm and willow.[4]

The types of tree identified in the charcoal fragments remind us of the importance of wood as a source of fuel for warmth and cooking. But the most important consequence of the discovery of the ice man is a fascinating demonstration of the subtle interactions between humans and trees over 5,000 years ago. It shows the great knowledge that humans had of the values and uses of different tree species and the enormous care that was taken to select the most appropriate species based on a wood's characteristics of strength, pliability and ease of working. The characteristic tree species in the area today, as it was 5,000 years ago, is spruce, although the variety of different species used by the ice man show that the wood was gathered from the transitional zone between the lower broadleaved woodlands, including species such as hazel and birch, and the higher spruce woods. Interestingly pollen from the hop horn-beam (*Ostrya carpinifolia*) was the dominant arboreal pollen found in the ice man's colon, and today this species remains one of the principal broadleaved trees in the lower Tyrolean valleys.[5]

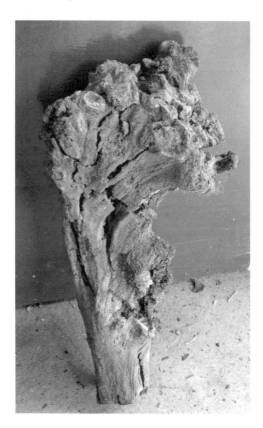

3 Ancient pollard
(dated at *c.* 2400 BC)
found at Aston/
Shardlow Gravel Pit,
Derbyshire, in 2002
by Chris Salisbury
and Norman Lewis.

Sweet Track and Flag Fen

The extent to which very large quantities of wood were used by our
Neolithic ancestors has recently been demonstrated by two major
archaeological research projects in lowland England. The Somerset
Levels is an area of low-lying land between the Mendip Hills to the
north and the Quantock Hills to the south. Since the last Ice Age the
area has been characterized by winter floods, and persistent waterlogging
has allowed the development of extensive peat bogs and the accumu-
lation of deep layers of peat several metres thick. Peat cutting for fuel
became common in the medieval period, and many drainage ditches
known locally as rynes were cut to help drain areas for agriculture. In
1834 a farmer who was cutting a ryne made a discovery that, although
largely ignored at the time, was later seen to be of great significance for
our understanding of the history of tree and woodland management.
The farmer uncovered deep in the peat a line of alder (*Alnus glutinosa*)
tree trunks that had been split and laid down parallel and next to each

other. When a later owner of the estate heard about this in 1864 he was so interested that he encouraged local archaeologists to consider the evidence. It was interpreted as part of an 'ancient plank road' or wooden trackway which was named the Abbot's Way, as it could have been constructed by one of the abbots of Glastonbury Abbey in the later Middle Ages to form a dry route over a wide stretch of boggy ground.[6]

In the 1930s and '40s further archaeological work uncovered various other trackways and with the development of radiocarbon dating it was realized that many of the wooden artefacts found in the peat originated from *c.* 3500 BC to *c.* AD 400 rather than the Middle Ages. One trackway, constructed of young rods of hazel, was dated to the third millenium BC, which gave 'the earliest evidence in the world for coppiced woodland', although this was not noted at the time.[7] By the 1960s digging for horticultural peat had become profitable, and this resulted in the discovery of more wooden archaeological structures. The Abbot's Way was rediscovered and traced for around 1,000 metres (3,280 feet). In 1970 Raymond Sweet found 'a piece of ash plank, clearly split from a large tree' and further digging found 'more of the same wood together with pegs still driven into the lower peats, and axed debris'. This structure, known as the Sweet Track, was soon identified as Neolithic and formed a remarkable cache of information about ancient woodland management.[8]

The aim of this Neolithic 'ingenious structure' was 'to provide a raised path across a wet reed swamp'. Long, straight trunks of quite thin ash, alder, hazel or elm were laid flat on the surface of the marsh. 'Pairs or groups of oblique pegs were then pushed or driven down into the soft unstable surface to either side' of this flat pole 'so that they crossed over it'. On 'the V formations thus created' planks were carefully balanced to make a strong walkway, and additional strength was provided by notches cut on the underside of the planks to fit them firmly to the structure (illus. 4). The types of wood used in the structure give an indication of the type of woodland growing in the Somerset Levels in the Neolithic period: 'oak, elm, lime and ash as the common large trees, hazel and holly for undergrowth, and alder, willow and poplar on the wetter fringes'. But the different pieces of wood found also give an indication of way the trees grew. For example, the lime trees that had been cut to produce long and straight planks were themselves tall, straight trees with few side branches, which suggested that they had grown fairly close together in dense woodland. Some of the large oak planks indicated that the oak trees used were up to 5 metres (16 feet 5 inches) long and 1 metre (3¼ feet) in diameter. Archaeologists were very impressed with the quality of woodworking skills indicated by the finds. Oak was the most frequent species used for planks and 'the trunks were converted into planks by splitting with wedges, either of stone or seasoned oak.' Most of 'the splits were radial, exploiting the tendency of oak to split along its rays' but smaller oak trunks were cut 'at right angles to the rays, more or less around the rings'.[9]

The large amount of preserved wood allowed archaeologists to carry out dendrochronological studies that produced a vivid picture of the types of Neolithic woodland. Large numbers of hazel rods were deliberately coppiced on a seven-year cycle. Moreover the age of the oaks used to make planks in the construction of the track ranged from 400 down to just over 100 years. This, together with the variety of species found, shows that by the late fourth millennium woodland was diverse and heavily influenced by human activities.[10]

Another major site of woodland archaeology, found at Flag Fen on the other side of England, just outside Peterborough, was excavated

4 A reconstruction of part of the 'Sweet Track', a Neolithic plank walkway in the Somerset Levels. Tree-ring evidence dates it to *c.* 3800 BC, and identifies the woods used as oak, hazel, ash and alder.

by Francis Pryor and a team of archaeologists from 1971 onwards. Peterborough had been designated a New Town and building development spread determinedly eastwards into the peaty fenland. Rescue archaeology revealed an extensive area of Bronze Age field systems with fields surrounded by ditches and a number of droveways. The 'Bronze Age fields were organized for the management of large numbers of livestock' and the droveways 'enabled animals to be moved between wetter and drier seasonally available grazing'. The fields were probably established in the later third and second millennium BC. Near Newark Road a complicated pattern of fields and other small enclosures and yards set about an important droveway indicated by substantial lines of post was uncovered. The very extensive quantity of wooden posts and planks, which have been variously interpreted as parts of a causeway and platform, allowed tree ring analysis to be carried out. This showed that most of the trees used for the various structures were felled just after 1300 BC and then, after a lull in activity, in a second batch between 1200 and 900 BC. The oldest posts found were of alder (*Alnus glutinosa*) and it is thought that these were felled locally. The other main species used was oak, which was also used for posts as well as planks, and a few wooden artefacts were found, including part of an axle and a wheel. The most common find was wooden poles, almost all of which had some evidence of Bronze Age woodworking, including the cutting of the posts to length, the removal of side branches, the sharpening of poles to make a pointed end and an occasional notch.[11]

The archaeologists at Flag Fen carried out a large number of experiments to establish the type of woodworking used in the Bronze Age. Splitting or cleaving was the principal means of making constructional timber. This is not to say that saws did not exist by this time; small copper saws had been developed in Egypt and Mesopotamia and the Minoans appear to have been the first to have saws, while Theophrastus (370–285 BC) argued that sawteeth should be set in alternate ways to allow the removal of sawdust, especially when green wood was being cut. This was reiterated by Pliny the Elder, who noted that the teeth of saws 'are bent each way in turn, so as to get rid of the sawdust', which would otherwise clog up the saw and make it ineffective. A fine Iron Age saw was found at the excavations at Glastonbury Lake Village in the early twentieth century. This saw 'is described as having its teeth "turned from side to side"; in other words, the teeth are like a modern saw, making this implement a transitional form between the simple serrated blade and the modern saw where the teeth are set so that they are effective on both the pull and the push.' It is curved with a finely cut knob at the end of the handle, which made pulling and pushing it easier.[12]

But cleaving was the most common way of making posts and other structural timbers at Flag Fen. The principle is to split each log length-ways in half, and then in half again, until the required thickness of timber is achieved. Experimentation showed that 'a log approximately 300 mm in diameter could be split in various ways depending on what was needed: half or quarter splits could make useful posts or beams, down to thirty-second splits, which will produce a stack of 32 feather-edge planks each about 150 mm wide with a thick edge of about 40 mm.' The archaeologist Maisie Taylor notes that by this method 'virtually no wood is wasted' with a tree 300 millimetres in diameter, but with larger trees, there is considerable waste as the halves and quarters are too large for normal use and require additional time-consuming work 'to bring them down to useful dimensions'. Following the Great Storm of 1987, experiments were made on the splitting of a large fallen oak from Minsmere, Suffolk. The importance of splitting the oak while it was still green and had not hardened through seasoning was soon identified, and the 'accuracy of the splitting' was affected by the careful positioning of different-sized wedges and the quality of the wood. The archaeologists were rather taken aback by the 'lifelike, almost heart-rending, noise' of 'a big tree' as it groaned while splitting and by the pungent 'tannic smell of the oak' after it was split 'sometimes pricking the eyes and back of the throat'. Their experimentation had almost taken them back directly into the sensory world of the Bronze Age.[13]

Large oak trees had a special significance at funerary and religious sites. At Foulmere Fen in Cambridgeshire a wooden structure inside a Neolithic barrow was found to have been built of tangentially cut planks 'virtually all made from one huge tree', carefully placed so that 'the out-side surfaces of the mortuary structure are also the outside of a tree, or trees.' Perhaps the most extraordinary archaeological oak from East Anglia, however, is the central oak from the Bronze Age timber circle on the shoreline at Holme-next-the-Sea in Norfolk. The timbers of the circle itself abut each other and are set with the bark to the outside of the circle. From a distance, therefore, 'the monument would have resem-bled a huge log or tree trunk'. The central tree was a substantial inverted oak tree more than a metre in diameter whose bark, unlike the surround-ing posts, had been removed deliberately. The function of this tree remains

5 Seahenge, the remains of a Bronze Age monument found below sea level at Holme-next-the-Sea, Norfolk. Tree-ring and radiocarbon evidence suggest a date of *c.* 2050 BC.

a mystery although it is thought that the roots of the upside-down tree may have 'cradled a body, perhaps left there during rites of excarnation'. Whatever its precise function, the inverted oak, stripped of its bark and surrounded by a massive 'trunk' of barked poles, demonstrates the potent interaction between prehistoric humans and trees.[14]

Classical knowledge

The younger Pliny (*c.* AD 61–*c.* 112), Gaius Plinius Caecilius Secundus, was a successful Roman senator and imperial administrator. Most wealthy Romans of his status owned several estates and in addition to his villa at Laurentum not far from Rome, he held two major estates at Tifernum on the Tiber (Città de Castello), about 150 miles northwest of Rome, and at Comum (Como) in northern Italy. His description of his Tifernum estate in a letter to his friend Domitius Apollinaris gives a fascinating insight into Roman woodland management. The estate is on the slopes of 'the Apennines, the most salubrious of mountains' and 'lies far back from the sea' and well away from the fever-ridden 'oppressive and noxious coast'.[15] He asked Domitius to 'Picture to yourself an enormous amphitheatre, such as only nature can provide . . . on the mountain tops are woods of great age, where the trees are tall. These provide hunting and in good measure and of great variety. On the slopes below are the coppice-woods; between the woods there are rich fields.' Lower down again the slopes 'are covered with vineyards in an unbroken pattern. Where the vineyards end, at the bottom of the slopes, plantations are growing up. Then come meadows . . .', and finally the heavy soils of the river plain. The younger Pliny demonstrates an acute awareness of the value of different types of land and woodland. This is emphasized in another letter, to Calvisius Rufus, asking for advice as to whether he should buy an estate adjoining and partly intermixed with his Tifernum estate where 'the land is fertile and rich and consists of arable fields, vines, and woods producing timber which provides a return that, though modest, can be depended on.' This regular, dependable income has been interpreted by the leading scholar of classical timber Russell Meiggs as a reference to coppicing.[16]

While we learn about the attitude of a Roman landowner to his woodland from the letters of the younger Pliny, it is his uncle Pliny the Elder (AD 23–79) and his encyclopaedic *Natural History* that provide us with insights into classical knowledge, experience and beliefs about trees. Pliny was born on the family estate at Como, which he left to his nephew, and had written his natural history by 77, a couple of years before his well-documented death from the eruption of Vesuvius in 79. The book was dedicated to his friend the Emperor Trajan's son Titus. Pliny organized his *Natural History* into 37 books: trees are found in books 12–16, between his descriptions of animals and agriculture. Books 12 and 13 discuss 'foreign trees that cannot be trained to grow elsewhere than in their place of origin and that refuse to be naturalized in strange countries'; book 14 focusses on growing vines and book 15 on fruit trees and olives; there is only a single book (16) on forest and timber trees grown in the Roman Empire.[17]

Pliny introduces trees by stating that 'trees and forests were supposed to be the supreme gift bestowed' by nature on mankind; they 'provided him with food; their foliage carpeted his cave and their bark served for raiment'. Trees were once 'the temples of the deities, and in conformity with primitive ritual simple country places even now dedicate a tree of exceptional height to a god'. He thought that people did not 'pay greater worship' to 'images shining with gold and silver than they did to forests and to the very silences that they contain'. He celebrates trees for their production of fruits, nuts and acorns, and especially because 'from trees are obtained olive oil to refresh the limbs and draughts of wine to restore the strength.'[18] The first tree that Pliny discusses is the plane, which fascinates him because it had been introduced to Italy and spread through the Roman Empire 'merely for the sake of shade'. He is thrilled by the celebrated plane tree growing in Lycia, southern Turkey, which 'stands by the roadside like a dwelling-house, with a hollow cavity inside it 81 feet across, forming with its summit a shady grove', while inside the tree embraced 'mossy pumice-stones in a circular rim of rock'. His contemporary Licinius Mucianus, who 'was three times consul' and governed Lycia, 'held a banquet with eighteen members of his retinue inside the tree', which 'provided couches of leafage' on such a 'bounteous scale' that 'he had then gone to bed in the same tree, shielded from every breath of wind, and receiving more delight from the agreeable sound of the rain dropping through the foliage' than he would from the 'gleaming marble' or 'gilded panelling' of a palace.[19]

Pliny loved a good story, and trees associated with the founding or history of Rome or with the spread of the Empire he found particularly attractive. He records the fig tree 'growing in the actual forum and meeting-place of Rome', which is 'worshipped as sacred', since it memorializes the fig tree 'under which the nurse of Romulus and Remus first sheltered these founders of the empire on the Lupercal Hill'. His antiquarian interests are drawn out when he recounts how the tree was named 'Ruminalis, because it was beneath it that the wolf was discovered giving her *rumis* (that was the old word for breast) to the infants'.[20] Although much of Pliny's work is based on his wide reading, some of his knowledge is derived from his own experience. He spent part of his life as a soldier in Germany and was enormously impressed by the 'vast expanse of the Hercynian oak forest, untouched by the ages and coeval with the world, which surpasses all marvels by its almost immortal destiny'. He was particularly struck by tales of the sea eroding the oak woods of the Netherlands around the Zuyder Zee: the oaks 'when undermined by the waves or overthrown by blasts of wind carry away with them vast islands of soil in the embrace of their roots'. These trees 'thus balanced, float along upright' so that the Roman fleets 'have often been terrified by the wide rigging of their huge branches when they seemed to be purposively driven by the waves against the bows of the ships at anchor for the night'. The Romans were then 'unavoidably compelled to engage in a naval battle with trees'.[21]

Russell Meiggs's survey of the classical literature on timber and trees in the ancient world shows that there was an enormous amount of interest in the cultural geography and history of trees and their products. But he was surprised to discover relatively little on what would later be termed forest science or silviculture, the planting of trees for growing timber or the cultivation and selling of trees. Indeed, he felt that 'One gets the impression from literature that woods were appreciated more for pasture and leaf-fodder than their timber.'[22] The dominant importance of knowledge about growing trees as part of normal farming and agriculture practices comes through the classical literature again and again. One of the most authoritative Roman texts on farming and estate management is *De agricultura* by Marcus Porcius Cato (234–149 BC), who had gained experience on the Sabian family estate as a young man. He very usefully lists the profitability of different types of farmland and woodland, with vines and irrigated gardens being the most profitable and woodland providing acorns and beech mast for pigs and cattle the least. Interestingly he names beds of willow coppice as more profitable than olive groves, and these were both more profitable than meadows and arable land for grain. Willows provided ties for vines, as they still do in many parts of Italy, and were useful for making all types of baskets and items of furniture. Coppice woods (*silva caedua*), which were used for firewood and many sorts of useful poles, and orchards lie between arable land and grazed woodland in terms of profitability.

Cato shows that the growing of trees was thoroughly integrated with the growing of vines and crops and he argued that trees should be managed carefully to maximize their potential for producing fruits, nuts, leaves and timber. It was expected that farms would grow their own trees, and he provides advice on the best way to cultivate elms, figs, pines and cypresses from seed. Farm buildings, including the farmhouse and outbuildings, would be built from timber grown on the farm. The qualities of different types of timber were well known: when constructing an olive press 'for the anchor-posts and guide-posts oak or pine are specified, for the great disc elm and hazel, because of their strength. Oak is used for dowels, cornel, one of the strongest woods, for nails, and willow for wedges. For the press-beam, black hornbeam should be used.'[23] The leaves of some trees are particularly valued as fodder for farm stock. Cato emphasized that elm and poplar provided excellent leaves for sheep and cattle. He thought that elm leaves were the best fodder, followed by poplar, and recommended that 'If you have poplar leaves mix them with the elm to make the latter last longer; and failing elms, feed oak and fig leaves.' He encouraged farmers to plant 'elms and poplars round the borders of the farm and along the roads to give you leaves for the sheep and cattle and timber when you need it'.

Marcus Terentius Varro, who had estates in Apulia and at Reate (Rieti in Lazio), writing in 37 BC, also thought that elms were one of the best trees to grow on a farm if the soil was suitable. He valued the elm especially as a tree over which to grow vines, while its leaves were excellent for sheep and cattle and its timber good for fencing rails and for firewood. Other authors had different ideas about the best tree on which to grow vines.[24] Lucius Junius Columella, after serving in the army, managed his Italian estates in Latium and Etruria. He, like Cato, thought that vines were the most profitable crop and that poplars, followed by elm and ash, were the best trees on which to grow them. These vine supports were carefully pruned, and the leaves, including ash leaves, were collected to be eaten by sheep and goats. Vines could also be supported by wooden stakes, which were usually cut from coppices. Columella argued that the best species for this purpose were oak, which took seven years to reach the correct size, and chestnut, which grew quicker and could be cut after five years. He thought that chestnut liked a 'dark, loose soil, does not mind a gravelly soil, provided that it is moist, or crumbling tufa; it is at its best on a shady and northward-facing slope'. Precise details are provided about planting and harvesting the chestnuts and the yield that could be expected, with every *jugerum* (0.25 hectare) yielding 12,000 stakes.[25]

Many classical descriptions of trees make sense to a modern reader and still hold true, but some essential distinctions that were believed for centuries now appear strange. One of the first people known to classify trees by considering their appearance and growth was Theophrastus, who was born on Lesbos and died at Athens. He was one of Aristotle's pupils and followed him as leader of the Lyceum. It is thought that Theophrastus' interest in the classification of plants was encouraged by the reports of geographers and botanists who had accompanied Alexander the Great's campaigns to the east. Theophrastus himself was not a great traveller, but drew evidence from friends who had visited areas such as Macedonia, Arcadia and Asia Minor. One of his basic assumptions was that, like animals, all trees were either female or male and that the former were fruit-bearing. He hit a problem when male and female trees of the same species both bore fruit, and argued that in this case the female trees had 'better and more abundant fruit'. The sex of the tree was also thought to affect the value of the timber. With lime trees, for instance, 'the wood of the male tree is hard, yellow, more fragrant and denser; the wood of the female is whiter'. Moreover the 'bark of the male is thicker and when it is stripped off it is hard and so does not bend, whereas the bark of the female is thinner and flexible'. This idea persisted through the classical period and neither Greek nor Roman authors recognized that 'most species of tree have male and female flowers on the same tree.'[26]

Trees were not only of value for farming and building; they were crucial for the construction of the navies on which the Greeks and Romans depended. Theophrastus gives one of the best descriptions of the types of timber used for different parts of the ships. He states that 'Fir (*elate*), mountain pine (*peuke*), and cedar (*kedros*) are the standard ship timbers.' The triremes and warships were built of 'fir because it is light', which made them faster and more efficient. Merchant ships, in contrast, were built of the heavier pine because it was less prone to rot. Theophrastus notes, however, that some states had to make do with the timber they had growing in their area. For example, in Syria and Phoenicia they used cedar because they had little pine or fir, while in Cyprus they used 'the coastal pine (*pitys*)' that grew on the island and seemed to be 'of better quality than mountain pine (*peuke*)'. The keel of triremes was made of oak because this was strong enough to withstand being hauled onto the shoreline, and they made 'the cutwater and catheads, which require special strength, of ash, mulberry, or elm'. The most favoured tree for the production of oars was the silver fir (*Abies alba*) and the best type were young, flexible trees that had been grown in fairly dense stands so that there were few side branches. Theophrastus pointed out that 'the fir has many layers, like the onion, for there is always a layer below the one that is visible.' He argued that it was important 'when they shave the wood to make oars they try to remove the layers one by one evenly'. If this was done successfully the oar would be strong, but 'if they do not strip off the layers evenly the oar is weak.'[27]

Homer's *Iliad* and *Odyssey* provide vivid insights into ancient appreciation and practical use of different trees.[28] The oak is the most commonly mentioned and figures in many similes, but there are also references to poplars, pines, firs and ash. In battle scenes the fall of warriors is likened to the felling of trees, the penetration of human flesh by bronze spears equated with the cutting of living trees with axes. In the *Iliad* the leader of the Cretans, Idomeneus, 'cast his spear on the throat below the chin' of Asius 'and drove the bronze clean through'. Asius 'fell as an oak falls, or a poplar, or a tall pine that among the mountains shipwrights fell with whetted axes to be a ship's timber'.[29] When Simoeisius was struck by Aias 'on the right of his chest beside the nipple' with a 'spear of bronze', he 'fell to the ground in the dust like a poplar tree that has grown up in the bottom land of a great marsh, smooth, but from its top grow branches: this a chariot-maker has felled with the gleaming iron so that he may bend a wheel rim for a beautiful chariot, and it lies drying by a river's banks'. Here the acute and precise picture of the poplar is followed by a poetic description of the way the tree is felled, seasoned and converted into chariot wheels.[30] The spearman Imbrius, who had married one of the daughters of Priam, was 'struck beneath the ear with a thrust of his long spear' by Ajax, and 'he fell like an ash tree that on the summit of a mountain that is seen from afar on every side is cut down by the bronze, and brings its tender leafage to the ground; so he fell.'[31]

Spears were usually made of ash with a bronze point. Hector struck Aias' 'ashen spear with his great sword close by the socket at the base of the point, and sheared it clean away' so that Aias was left brandishing 'vainly a pointless spear, and far from him the head of bronze fell to the ground with a clang'. Achilles' spear was the 'Pelian spear of ash, that Cheiron had given to his dear father from the peak of Pelion (near present-day Lamia), to be for the slaying of warriors'.[32] The noise of battle is equated with wind and fire rushing through the forest. When the Trojans and Achaeans raised 'a terrible shout as they leapt on each other' Homer compares this to 'the roar of blazing fire in the glades of a mountain when it leaps to burn the forest' and 'the shriek' of the wind 'among the high crests of the oaks – the wind that roars the loudest in its rage'.[33] The shouts and calls of the two 'masters of the war cry' Hector and Patroclus, and the sounds of the Trojans and Achaeans slaughtering each other, are compared to the noise of the 'East Wind and the South' as they 'strive with one another in shaking a deep wood in the glades of a mountain – a wood of beech and ash and smooth barked cornel' which 'dash one against the other their long boughs with a wondrous din, and there is a crack of broken branches'.[34] But Homer was also aware that trees that survive frequent strong winds have usually developed strong roots. When Polypoetes and Leonteus defended a gate against King Asius, they 'stood firm like oaks of lofty crest on the mountains that ever stand up to the wind and rain day by day, firm fixed with roots great and long'.[35]

The day to day life of woodcutters is glimpsed through analogy. In one battle Agamemnon and the Danaans broke through the opposing line 'at the hour when a woodman makes ready for his meal in the glades of a mountain, when his arms have grown tired with felling tall trees, and weariness comes on his heart, and desire of sweet food seizes his thought'.[36] But the great care, skill and attention needed when felling trees with an axe is recognized by Nestor when he tells his son Antilochus, in a pep talk before a chariot match, that a charioteer, like a woodman, needs intelligence over brawn: 'By cunning, you know, is a woodman far better than by might.'[37] The use of ancient Greek woodland for keeping pigs and hunting wild boar is also evident. In the *Odyssey* Circe's palace was built in 'the forest glades' where there were 'mountain wolves and lions'. She drugged the companions of Odysseus and 'penned them in sties'. They had 'the heads, and voice, and bristles, and shape of swine, but their minds remained unchanged even as before'. As they wept in their pen 'before them Circe flung mast and acorns, and the fruit of the cornel tree, to eat, such things as wallowing swine are wont to feed upon.'[38] In the *Iliad* two Trojans fought 'like a pair of wild boars that among the mountains await the tumultuous throng of men and dogs that comes against them'. They charge at each other and 'crush the trees about them, cutting them at the root, and there arises the sound of the clash of tusks till someone strikes them and takes away their life'.[39]

Although most of the dendrological similes in Homer concern bloody battle scenes, there are exceptions which bring out subtle ways of relishing the beauty of trees. In the *Odyssey* the women who work in the house of Alcinous 'weave at looms or twist the yarn, while, like the leaves of a tall poplar, flit the glancing shuttles through their fingertips'.[40] The Achaens feasted on freshly slaughtered and cooked cattle and offered gifts 'to the immortals on the holy altars, beneath a fair plane tree from which flowed the bright water', although here the beauty is ironical, since 'a serpent, blood-red on its back, terrible' then 'glided from beneath the altar and darted to the plane tree' where it devoured the 'nestlings of a sparrow' that cowered beneath the leaves 'on the topmost bough'. Here the beauty of the tree accentuates the horror of the arrival of the terrible serpent sent by Zeus.[41]

Even the greatest warriors could succumb to the beauty of plane trees: when the Persian Emperor Xerxes was travelling between Phrygia and Sardis on the road which crossed 'the river Maeander' in 480 BC he found a plane tree under which he camped overnight and 'which he adorned with gold because of its beauty, and he assigned one of his immortals to guard it'.[42] By the first century AD the plane had become one of the most popular and accepted shade trees in the Roman Empire. It was used to shade those engaged in education, athletics and training for war. Pausanias describes how at Sparta in the second century AD there was an area called 'the Plane-tree Grove, so called from the plane-trees which grow in an unbroken line around it', which was where 'young Spartans passing from adolescence to manhood' did their rough fighting: 'they strike, and kick, and bite, and gouge out each other's eyes. Thus they fight man against man.' Before the fight started 'the lads pit tame boars against each other, and the side whose boar wins generally conquers in Plane-tree Grove.' He also describes how, at the 'old gymnasium' at Elis, 'high plane trees grow between the tracks', shading the area where the athletes did 'the training through which they must pass' before the Olympic Games.[43] And the plane, which became one of the most popular trees introduced from Greece to Italy, was one of the younger Pliny's favourite trees, although, rather than celebrating an individual tree like Xerxes, he cultivated and marshalled hundreds of them in the extensive formal gardens at his Tuscan villa. One of his favourite places, which he used when he had 'none but intimate friends with me', was a 'summer-house enclosing a small area shaded by four plane-trees, in the midst of which rises a marble fountain which gently plays upon the roots of the plane-trees'. He also had a hippodrome, which was horseshoe-shaped and used for walking as well as riding. The straight rides were shaded and 'set round with plane-trees covered with ivy, so that, while their tops flourish with their own green, towards the roots their verdure is borrowed from the ivy that twines round the trunk and branches, spreads from tree to tree, and connects them together'. At the curved ends of the hippodrome, cypresses cast a 'deeper and gloomier shade' than the plane trees; additional dwarf planes were to be found in an adjoining garden, with fruit trees, box trees and laurels.[44] Pliny the Younger's descriptions of his gardens depict the Roman use of trees at its most luxurious yet productive.

But the Roman enthusiasm for trees and certain timbers could be destructive, as is shown by the fate of one of the most fashionable timbers in the Roman Empire, the mysteriously attractive wood of the 'citrus tree' *Tetraclinius articulata*. This evergreen coniferous tree was famed for its timber, in contrast to the trees with the same common name in the genus *Citrus*, which provide lemons and oranges. Some Roman men became completely infatuated with tables made from citrus wood and

Pliny remarked that 'ladies use [men's] table-mania' as a useful 'retort' when charged with 'extravagance in pearls'. The mania for these tables was as strong if not stronger than the tulipomania which beset the Netherlands in the seventeenth century. Cicero, who was not a particularly wealthy man, purchased one for 500,000 sesterces and one owned by Gallus Asinius 'cost a million'; these, according to Pliny the Elder, were sums of money equivalent to the price of a reasonable estate, 'supposing', he added acidly, 'someone preferred to devote so large a sum to the purchase of landed property.' The main attraction of the citrus wood tables were the 'wavy marks forming a vein or else little spirals. The former marking produces a longish pattern and is consequently called tiger-wood, while the latter gives a twisted pattern and consequently slabs of that sort are called panther-tables.' Other sorts of wood, depending on the markings, looked like peacocks' tails or parsley. The 'highest value of all resides in the colour', and highly polished citrus wood tables had the enormous advantage of not being 'damaged by spilt wine, as having being created for the purpose of wine-tables'.[45]

The demand for citrus wood tables was intense and people collected them: Seneca is said to have amassed 500 with ivory legs. This huge demand caused the destruction of areas of woodland where the trees grew in North Africa. The best citrus wood was cut from the roots of the tree and the patterning was a type of burr wood, which, as Pliny noted, was caused by 'a disease of the trees' found growing as an 'excrescence of the root' or 'as knobs that grow above ground, on branches as well as on the trunk'. Strabo noted that Mauretania was 'surpassing in the size and in the number of its trees' and was the country 'which supplies the Romans with the tables that are made from one single piece of wood, very large and most variegated'. Other types of wood such as maple were used as a substitute and Strabo noted that some trees in the Ligurian forests around Genoa had 'very great quantities' of 'timber that is suitable for shipbuilding, with trees so large that the diameter of their thickness is sometimes found to be eight feet. And many of these trees, even in the variegation of their grain, are not inferior to the thyine wood [citrus] for the purposes of table making.' Pliny records that the 'most celebrated citrus-wood' came from a 'mountain called Ancorarius' in Mauretania but that 'the supply is now exhausted.' The full effect of this exploitation was depicted carefully by Lucan, who noted that the 'timber of Mauretania' had been 'the people's only wealth' but that they were 'content with the leafy shade of the citrus-tree', being 'ignorant how to make use of this wealth'. But 'our axes have invaded the unknown forest and we have sought tables as well as dainties from the end of the earth.' And thus Roman greed for luscious, highly patterned trees resulted in the destruction of the citrus wood forests, a portent of the removal of huge areas of woodland and its conversion to meadows, pastures, fields and deserts over the centuries that followed.[46]

TWO
Forests and Spectacle

KINGS, bishops, aristocrats and state authorities have all had a particular interest in controlling their forests. Forests were defined to impose order over large areas of land and as a way of policing borders. The spectacle of hunting and its association with military prowess and horsemanship were key to the establishment of power by Norman kings in England, as they had been for Alexander the Great in Macedonia and Roman emperors such as Hadrian. In Europe forest laws were established in the early Middle Ages to control large areas of land that included villages and even towns. Literary depictions of forests drew on associated and conflicting ideas of chaos, freedom, contemplative spaces and danger, and the strength of these ideas had a direct effect on the way that forests were administered, controlled and managed for centuries. Great kings were keen to memorialize their prowess at hunting and control of their forests. Nebuchadnezzar, the powerful young king of Babylon (r. 605–562 BC), celebrated his victories against the Phoenicians and attack on Jerusalem with an inscription at Wadi Brisa in present-day Lebanon. The text 'is accompanied by a relief, now very badly worn, of the king killing a lion', and a second relief shows 'the king cutting down a tree'. Nebuchadnezzar's death was celebrated by the prophet Isaiah: 'the whole world has rest and is at peace . . . The pines themselves and the cedars of Lebanon exult over you. Since you have been laid low, they say, no man comes to fell us' (14:7–8).[1]

In 1977 three royal tombs were discovered at the modern village of Vergina in northeastern Greece.[2] Among the many astonishing finds was a large tomb interpreted as being that of Philip II, father of Alexander the Great, who had been murdered at the age of 46 by his ex-boyfriend in 336 BC. Above Philip's tomb was the funeral pyre which contained remnants of the hounds and horses which hunted with him.

The attribution remains contested, although if not Philip it is likely to be that of Arrhidaeus (later Philip III), Alexander the Great's brother. The doorway of the tomb is between two Doric columns and above these is an Ionic frieze showing a hunting scene. James Davidson has described it as 'a vigorous hunt in a sacred landscape with naked youths and dogs attacking deer, a boar and a bear, with spears, a net and an axe'.

The scene is remarkable in many ways, but for many the most exciting element is the central group around the lion. Here a bearded man on a rearing horse strikes downwards to kill the lion while another rider in the centre of the frieze is in the act of throwing a javelin with the same aim. The bearded man 'is supposed to be Philip making use of the king's prerogative of killing the king of beasts, while his son, the youthful Alexander, rushes to assist him'. The young men 'are the so-called Royal Boys who alone were allowed to accompany the king on hunts like these'. With all this to catch the eye, it is easy to miss the landscape in which the hunt is taking place. Although the original frieze is faded, the ground is fairly clear of vegetation, though rocky in places, and appears to be grazed by sheep or goats; such open areas were essential for hunting on horseback. There are several old trees which resemble pollards, trees that have been regularly cropped of their branches; one in the centre next to Alexander's horse Bucephalus, if the attribution is correct, appears to be dead. This ancient landscape is reminiscent of many equivalent modern landscapes of old trees in pasture and heath.

6 Doorway to the tomb of Philip II, *c.* 336 BC, Vergina, Greece.

Alexander's love of hunting was described by the Roman senator Quintus Curtius Rufus in his biographical account of Alexander fighting deep in Asia. There were 'no greater indications of the wealth of the barbarians in those regions than their herds of noble wild beasts, confined in great woods and parks'. They chose for this purpose 'extensive forests made attractive by perennial springs' and they 'surround the woods with walls and have towers as stands for hunters'. When Alexander arrived, 'the forest was known to have been undisturbed for four successive generations' and 'entering it with his whole army' he 'ordered an attack on the wild beasts from every side'. A 'lion of extraordinary size rushed to attack the king himself' and Lysimachus, who happened to be next to Alexander, prepared to attack it with a hunting spear. But Alexander 'pushed him aside and ordered him to retire', 'met the wild beast' and killed it with a single thrust of his weapon.[3]

Hunting was seen by the Romans 'as a spectacle to entertain the people' and was popular with the Emperor Trajan, whose 'only relaxation', according to Pliny, was 'to range through the forests and drive the wild beasts from their lairs' while Pliny himself 'sat by the hunting-nets, with writing materials instead of hunting spears'. Trajan's successor Hadrian was particularly fond of hunting. It is probable that he hunted wild boar on his visit in AD 122 to the northernmost reach of his empire where Hadrian's Wall was built: an altar in the northern Pennines records that a Roman officer 'had taken a boar of exceptional fineness which many of his predecessors had been unable to bag'. In 124 he travelled through present-day northwest Turkey and hunted in the wooded interior, founding a town named Hadrianutherae ('Hadrian's Hunts', now Balıkesir) after a successful hunt resulting in the death of a female bear. Coins were minted which showed the Emperor Hadrian on horseback with the head of the bear on the other side. The inscription on the tomb of one of Hadrian's favourite hunting horses, Borysthenes, described how he 'used to fly over plains and marshes and hills and thickets, at Pannonian boars – nor did any boar, with tusks foaming white, dare to harm him'.[4]

An Egyptian hunting trip Hadrian took with Antinous in September 130 is described in a poem by Pankrates which survives in Greek on a fragment of papyrus. Hadrian 'wished to test the aim of the handsome Antinous' and deliberately wounded a lion which 'grew even fiercer and tore at the ground with his paws in his rage . . . he lunged at them both in rage.' Eventually the lion mauled Antinous' horse and Hadrian

7 *Boy with Horse*, AD 117–18, marble relief. This bas-relief was excavated by Gavin Hamilton in part of Hadrian's villa at Tivoli. It possibly represents Castor taming his horse.

killed the lion, saving Antinous' life.[5] Hadrian was 'a passionate, almost obsessive hunter' who composed epigrams extolling the virtues of favourite hunting dogs and horses. There was a relief of a young man with a horse at Hadrian's Tivoli villa and there are eight marble relief tondi celebrating his hunting exploits and his huntsmen, including a boar and a lion hunt. The 'figure in the background of the boar hunt tondo, riding behind Hadrian, bears a strong resemblance to Antinous'.[6]

The central place of hunting and trees within human culture can be traced back even further. In 2010 it was claimed that an archaeological team had discovered the remains of a stone enclosure, possibly dating from the twelfth and thirteenth centuries BC, near Nemi in the Alban Hills south of Rome. This enclosure 'amidst the ruins of an immense sanctuary dedicated to Diana, the goddess of hunting, along with the remains of terracing, fountains, cisterns and a nymphaeum, once surrounded a large sacred tree, such as the one that the pre-Roman Latins believed symbolized the power of their priest-king.' Christopher Smith, director of the British School at Rome, commented that 'this is an intriguing discovery and adds evidence to the fact that this was an extraordinarily important sanctuary; we know that trees were grown in containers at temple sites and that the Latins gathered here to worship right up until the founding of the Roman republic in 509 BC.'[7] This could provide a location for the famous scene with which Sir James Frazer opens his monumental work on the anthropology of religion, *The Golden Bough* (1890).

Frazer ruminates over the landscape around Lake Nemi where 'Diana herself might still linger by this lonely shore, still haunt these woodlands wild. In antiquity this sylvan landscape was the scene of a strange recurring tragedy.' Under the steep cliffs below Nemi, on the shore of the lake, 'stood the sacred grove and sanctuary of Diana Nemorensis, or Diana of the Wood' and in this wood 'there grew a certain tree round which at any time of the day, and probably far into the night, a grim figure might be seen to prowl.' Frazer pictures how this figure would 'darken the fair landscape' for the 'gentle and pious pilgrims' at Diana's shrine. A belated wayfarer would see a sombre scene: 'the background of forest showing black and jagged against a lowering and stormy sky, the sighing of the wind in the branches, the rustle of withered leaves under foot . . .'. This was the priest of the sanctuary, who always carried a drawn sword and might expect to murder or be murdered at any time, for the rule of the sanctuary was that the 'man for whom he looked was sooner or later to murder him and hold the priesthood in his stead'. Within Diana's sanctuary at Nemi grew a tree from which no branch might be broken (illus. 67). The only exception was that 'a runaway slave was allowed to break off, if he could, one of its boughs.' If he succeeded in this task he had the right to fight the 'priest in single combat, and if he slew him he reigned in his stead with the title of King of the Wood (*Rex Nemorensis*)'. Frazer argues that the 'public opinion of the Ancients' was that 'the fateful branch was that Golden Bough' which 'Aeneas plucked before he essayed the perilous journey to the world of the dead.' Moreover the golden bough was actually mistletoe, which was seen as the 'the life of the oak' since it remained 'green while the oak itself is leafless'.[8]

When Aeneas searches for the perfection of the spirit, represented by the golden bough, forests signify dangerous, shadowy and dark places where 'lust and unbridled passion' rule.[9] Aeneas and his followers are commanded by the Sibyl to make a pyre for Misenus' tomb and go to the 'forest primeval, the deep lairs of beasts; down drop the pitchy pines, and ilex rings to the stroke of the axe; ashen logs and splintering oak are cleft with wedges, and from the mountains they roll

8 William Ottley, *Study of Trees and Rocks in the Chigi Park at Ariccia*, 1790s, pen and brown ink drawing, with brown-grey wash.

9 *The Golden Bough*, *c.* 1847, etching by Thomas Abel Prior after Joseph Mallord William Turner.

in huge rowans'. Alone, Aeneas 'ponders with his own sad heart, gazing on the boundless forest' and goes on to find the golden bough and the way to the Underworld: 'As in the winter's cold, amid the woods, the mistletoe, sown of an alien tree, is want to bloom with strange leafage, and with yellow fruit embrace the shapely stems: such was the vision of the leafy gold on the shadowy ilex.' Aeneas immediately 'plucks it and greedily breaks off the clinging bough, and carries it' to the Sibyl.[10]

The *Aeneid* was an enormously influential text in the medieval period and its symbolic forest landscapes are associated with hunting, fighting, exile and death. When the Trojans landed in Libya, for example, above the harbour loomed 'heavenward huge cliffs and twin peaks' and 'a background of shimmering woods with an overhanging grove, black with gloomy shade'. The first thing Aeneas does is to climb a peak to look for other ships, but instead he sees 'three stags' with herds of deer and he immediately seizes 'his bow and swift arrows' and 'lays low the leaders themselves, their heads held high with branching antlers, then routs the herd and all the common sort, driving them with his darts amid the leafy woods'.[11] In Book IX, when the warriors Euryalus and Nisus are chased, they flee 'to the wood and trust to the night' but they

10 *Dido and Aeneas*, 1787, etching and engraving by Francesco Bartolozzi (figures) and William Woollett (landscape), after Thomas Jones (landscape) and John Hamilton Mortimer (figures).

find the 'forest spread wide with shaggy thickets and dark ilex; dense briers filled it on every side', and although 'here and there glimmered the path through the hidden glades', Euryalus 'is hampered by the shadowy branches'. Nisus gets through the wood, but calls out 'Unhappy Euryalus, where have I left thee? Or where shall I follow, again unthreading all the tangled path of the treacherous wood?'; after much bloodshed both warriors are killed in the forest (illus. 11).[12]

The forest in the *Aeneid* 'appears as a landscape of potentiality, associated with destiny, prophecy and the unexpected' and this can be contrasted with the harmony of idyllic pastoral groves depicted in Virgil's *Eclogues*.[13] The First Eclogue opens with Tityrus, who may represent Virgil, 'at ease beneath the shade' of 'the spreading beech's covert', telling the shepherd Meliboeus, who has been driving his goats 'amid the thick hazels', of his visit to Rome, which has 'reared her head as high among all other cities as cypresses oft do among the bending osiers'.[14] In a later Eclogue Meliboeus listens from the shade of a 'whispering ilex' and near a 'hallowed oak', where 'swarm humming bees' to a poetry competition where Corydon tells of 'junipers and shaggy chestnuts; strewn about under 'e trees lie their own diverse fruits; now all nature smiles'; while Thyrsis considers 'Fairest is the ash in the woods, the pine in the gardens, the poplar by the rivers, the fir on the mountain-tops.'[15]

One of the most significant trees from a religious perspective is the tree of knowledge of good and evil in the Garden of Eden, the home of Adam and Eve in the Book of Genesis (illus. 12). The Garden of Eden is popularly called Paradise, a word which 'is probably of Persian origin and signified originally a royal park or pleasure ground'. The *Catholic Encyclopaedia* notes that the word paradise 'does not occur in the Latin of the Classic period nor in the Greek writers prior to the time of Xenophon' and in the Old Testament 'it is found only in the later Hebrew writings in the form (*Pardês*).' An example of the origin of the term is provided when Nehemiah, a high official in the Persian court of King Artaxerxes I, returned to Jerusalem in 445 BC to repair the city:

> Moreover I said unto the king: 'If it please the king, let letters
> be given me to the governors beyond the River, that they may
> let me pass through till I come unto Judah; and a letter unto
> Asaph the keeper of the king's park, that he may give me timber
> to make beams for the gates of the castle which appertaineth to

Me, me; adsum qui feci; in me convertite ferrum,
O Rutuli....

Vieira Portuensis Delin.   F. Bartolozzi R.A. Sculp.

11 *Death of Euryalus*, 1800, etching and engraving by Francesco Bartolozzi after
Francisco Vieira Portuense.

the house, and for the wall of the city, and for the house that I shall enter into.'

Here the word *happerdês* is used to describe Asaph's role as the custodian of the royal park of the Persian ruler (Nehemiah 2:7–8).

According to Genesis, man is 'set to take care of the Garden of Eden' and has 'permission to eat of its fruit, except that of the tree of the knowledge of good and evil'. Adam and Eve 'live in childlike innocence until Eve is tempted by the serpent, and they both partake of the forbidden fruit'. They then know sin, 'incur the displeasure of Yahweh' and are thrown out of the Garden of Eden. Consequently 'their lot is to be one of pain and hardship', and humans have to win 'sustenance from a soil which . . . has been cursed with barrenness'.[16] Trees are also of crucial significance in the Norse myths, including the creation of humankind. The rich Icelandic landowner, lawyer and poet Snorri Sturluson (1179–1241) compiled the important collection of oral myths and poems known as the Prose Edda. In the myth concerning the origin of humankind Snorri tells how

> Bor's sons were walking by the sea-shore and came upon two logs. They picked them up and shaped them into human beings. The first gave them breath and life, the second understanding and motion, the third form, speech, hearing and sight. They

12 John Martin, *Fall of Man*, 1831, mezzotint with etching.

13 School of
Annibale Carracci,
*Garden of Eden
with Eve and the
Serpent*, late 16th
century, ink
drawing.

gave them clothes and names. The man was called Ask [ash tree], the woman Embla [perhaps 'elm' or 'vine']. From them descend the races of men who have been given a dwelling-place below Midgard.[17]

When the logs became human 'The sun shone from the south on the stones of earth; then the ground was grown with green shoots.'[18]

After this, Snorri tells of 'the Yggdrasill, tree of fate, upon which the welfare of the universe seems to depend. Beneath it lay the well of fate (*Urðarbrunnr*) from which the fates, conceived in female form, proceeded to lay down the course of men's lives.'[19] This tree is usually identified as an ash tree. But 'even the great world-tree is subject to attack':

The ash Yggdrasill endures hardship
More than men can know,
The hart bites its crown, its sides decay,
The serpent Nidhogg tears its roots.

The Norns tried to preserve the ash Yggdrasill 'by pouring over its branches water and mud from the Well of Fate. This magic liquid helps to stop the rot. In the end the tree is to fall, as are the gods themselves. They are as mortal as man.'[20]

Trees were associated with several Norse myths and in the *Hávamál* 'it is told how Òðinn hung for nine nights on a windswept tree':

I know that I hung
on the windswept tree
for nine full nights,
wounded with a spear
and given to Òðinn
myself to myself;
on that tree
of which none know
from what roots it rises.

14 Albrecht Dürer, *Adam and Eve; standing on either side of the Tree of Knowledge with the Serpent*, 1504, engraving.

15 Hans Varnier the Elder, *The Tree of Knowledge with the Serpent*, 1530s–40s, woodcut.

Many have equated this tree with a pagan reflection of Christ on the Cross and the 'similarities between the scene described here and that on Calvary are undeniable'. Indeed 'the two scenes resembled each other so closely that they came to be confused in popular tradition.' But most elements in the Norse myth can be explained as part of the pagan tradition.[21] According to the eleventh-century scholar Adam of Bremen, a 'notorious festival at Uppsala was held every nine years, and continued for nine days. Nine head of every living thing was sacrificed, and the bodies were hung on trees surrounding the temple.' These hanged victims could well be 'dedicated to Òðinn, whose image stood with those of Thór and Fricco in the temple of Uppsala', and scholars consider that 'the tree from which Òðinn swung was no ordinary tree. It can hardly be other than the World Tree, the holy Yggdrasill', named after Òðinn's horse, from 'Yggr' (the terrifier) and 'drassil' (a poetic word for horse).[22]

One troublesome detail is that the Yggdrasill is sometimes described as 'evergreen', and some argue that the tree of life could be a yew rather than an ash. There is 'little doubt that the evergreen yew was held sacred, whether or not the Yggdrasill and holiest tree at Uppsalir were yews. The best bows were made from yew.' Moreover, the archer god Ull, also known as the hunting god, who was 'so skilled on skis that none could compete with him', lived in Ýdalir (Yew-dales), 'where yews flourished'. Adam of Bremen described the temple at Uppsala in *c.* 1070 when it was still in use: it was 'as if centuries of heathen belief and practice had silted up in this Swedish backwater'. In the grove next to the temple itself

> One tree was holier than all the others; it was evergreen
> like the ash Yggdrasill. In some ways it resembled the great
> column, Irminsul which, as Saxons believed, upheld the
> universe. We may also think of Glasir, the grove with golden
> foliage standing before the doors of Valhöll, and of the tree
> growing from unknown roots on which Òðinn swung in his
> death agony.[23]

Biblical and classical conceptions of hunting and forests strongly influenced the writers of the Middle Ages, who linked ideas of the wilderness and the desert and classical ideas of the chaotic forest with the uncultivated landscape of the medieval forest. Associations

16 The world-tree
Yggdrasill from
the Icelandic *Edda
oblongata* manuscript
of 1680.

of solitude and divine inspiration 'were appropriated as part of the forest's symbolism in the romances', which became one of the most popular forms of literary entertainment at the courts of the north European aristocracy. Deserts and forests were both characterized by their uncultivated quality. Biblical deserts often consisted of rocky landscapes with caves, some vegetation and springs. For example in St Athanasius' *Life of St Anthony*, written in AD *c.* 360, the Egyptian deserts included a cave with seedlings, trees and wild beasts.[24] This 'exchange of landscapes' preceded the Norman Conquest of England: Aelfric (*c.* 955–1005), the abbot of Eynsham, referred to John the Baptist going to the wilderness 'to escape the vices which men practise', where 'He drank neither wine, nor beer, nor ale, nor any of the drinks which men drink. But he ate fruit and that which he could find in the forest.' Solitary ascetics restricted their existence by staying in caves or the hollow interiors of ancient trees. The 'associations of the forest with the Biblical desert or wilderness render comprehensible such stories as that of Evrard de Breuteuil, viscount of Chartres, who abandoned all to lead an eremitic life as a charcoal-burner in 1073, modelling himself after Saint Thibaud'.[25]

In Anglo-Saxon and Norman times felons were 'put outside the law' and were said to bear 'the wolf's head' and to be 'treated as wolves by those within the law'. An early fourteenth-century Anglo-Norman poem tells of a guiltless person accused of a felony whose only remedy was to go to 'the beautiful shade' of the wood of Belregard; 'There is no deceit there, nor any bad law . . . where the jay flies and the nightingale always sings without ceasing.'[26] Woods and forests played a complex and central role in medieval romance landscapes. In influential romance texts such as those by the twelfth-century writer Chrétien de Troyes, who first wrote of that exemplar of chivalry Sir Lancelot, the departure of the knight on solitary quests and adventures through the unknown forest is a strong theme.

At the court of King Arthur, for example, the knight Yvain flees the tournament at Chester when he is spurned by Lady Laudine. 'He would rather be banished alone in some wild land, where no one would know where to seek for him, and where no man or woman would know of his whereabouts.' He rose 'from his place among the knights' and 'such a storm broke loose in his brain that he loses his senses; he tears his flesh and, stripping off his clothes, he flees across the meadows and fields.' He meets 'close by a park a lad who had in his hand a bow and five barbed

17 Geertgen tot
Sint Jans, *John
the Baptist in
the Wilderness,*
c. 1490–95, panel.

arrows, which were very sharp and broad' but 'he had no recollection
of anything that he had done. He lies in wait for the beasts in the woods,
killing them, and then eating the venison raw. Thus he dwelt in the
forest like a madman or a savage.'[27] Forests were to a large extent 'the
special territory of romanciers', who in the later medieval period 'looked
back to a time immediately preceding their own, largely mythical in
England, in which this forest landscape was almost boundless, and
thus might be moulded to the requirements of the romance form'.[28] The
forest was a place of exile where one's will, courage and expertise could
be tested to the full. It was also a place where those outside the law could
go to hide from society.

Practical spectacle: medieval Royal Forests

We know that the forests of England were not boundless in the medieval period; on the contrary, they were defined by their very boundedness through the application of special laws. Areas of land, including dense woodland, but also open areas of heath, moor and pasture, were protected for hunting by Frankish kingdoms by as early as the seventh century, and 'England maintained close contact with these kingdoms and rapidly assimilated their cultural tradition.' Some English estates, such as Bickleigh in a pre-Conquest charter of AD 904, 'were singled out as of special importance for hunting'. The landscape historian Della Hooke's close examination of Anglo-Saxon charters has allowed her to identify the Old English word *haga* as describing 'wood-banks topped by hedges or fences' and that '*haga* features noted in woodland, therefore, appear to have been enclosures into which deer could be encouraged as a readily available source of venison and protected from marauding animals such as wolves.'[29]

But it was the arrival of the Normans with their new legal systems and vast enthusiasm for hunting that grounded the complex bundle of ideas associated with forests and hunting into a hardnosed reality in England and Wales. When William, duke of Normandy (1027/8–1087), conquered England in 1066, hunting, which had been popular with the Anglo-Saxons, was reaffirmed as a vitally important royal sport by the creation of large areas of royal forest, including the New Forest, where William's son Richard was killed in a hunting accident. William was 'physically imposing' and 'was exceptionally strong; William of Malmesbury, for example, recounts that, while spurring on a horse, he could draw a bow which other men could not even bend.' Hunting as well as demonstrating the power and authority of the monarch was essential training and practice for warfare. In 1069 a Danish army landed in the North and was 'joined by Edgar Ætheling and a large force of English rebels, and the combined army captured York on 20 September'. It is thought that William heard of the loss of York while hunting in the Forest of Dean.[30]

William II (*c.* 1060–1100), known as William Rufus, who succeeded William I in 1087, was a 'rumbustious, devil-may-care soldier, without natural dignity or social graces, with no cultivated tastes', but his 'chivalrous virtues and achievements were all too obvious' and he 'maintained good order and satisfactory justice in England and restored good peace to Normandy'. He was also exceptionally keen on hunting and his 'enlargement of the royal hunting preserves' and 'tightening of the severity of the laws to protect them were generally resented'. His last hunting trip was in the New Forest on Thursday, 2 August 1100. The evidence suggests that 'Contrary to his usual custom he did not go out until the afternoon, when the royal party, which included his brother Henry, broke up into small groups, each to take up position at a butt.' Here they dismounted and waited 'to shoot at deer driven across their front by beaters'. Disastrously, it was not a deer that was shot, but the king, who was 'killed instantly by an arrow in the heart'. The exact site is disputed but it is likely to be near where the 'Rufus Stone' was set up in 1745. It was most likely a hunting accident: William's elder brother Richard had been killed in the same way, as had one of his nephews. But several anti-Norman writers argued that the archer who fired the arrow

The DEATH of WILLIAM RUFUS.

18 *The Death of William Rufus*, 1777, etching and engraving by Francis Chesham after John James Barralet.

'was simply God's instrument to avenge the making of the New Forest and to punish a blasphemer'.[31] Hunting was a very dangerous sport, as were the celebrations after a day's hunting. In November 1135 William Rufus's younger brother and successor Henry I (1068/9–1135) travelled to his hunting lodge at Lyons-la-Forêt in Normandy 'to indulge in his favourite pastime of hunting' but 'fell mortally ill' after 'feasting on lampreys – a delicacy that his physician had forbidden him'.[32]

The three kings of England from 1066 to 1135 appeared almost to live and die for hunting, and by the reign of Henry I the royal forests were controlled by 'an administrative system fully developed and functioning in a routine manner'. There is 'evidence that each of the three kings manipulated the forest administration as a matter of conscious policy when it could be to his advantage'. Moreover, in the subsequent civil war both sides used 'grants of exemption from the forest' in attempts to gain support. After the war Henry II (1133–1189), who became king in 1154, 'extended the royal forests beyond the area they had attended under his grandfather' and they covered their greatest area during his long reign.[33] Henry II was as keen on hunting as his Norman ancestors and also engaged in 'intellectual debates with a circle of clerks or visiting monks', but at 'moments of tumult at court he fled in silence to his beloved forests, seeking a solitary peace in the wild'. He improved the governance of the forests and 'the assize of the forest (1184) brought the regulation of forest offences, previously based largely on the king's whim, into the realm of customary law.'[34]

But the increase in the area of forests was the cause of resentment and disputes with landowners, which came to a head in the reign of King John and brought about the Forest Charter of 1217. This resolved several contentious issues and the rising tide of new forests was stopped and pushed back: new forests established by Richard I and John were disafforested and the boundaries of those created by Henry II were checked. Woods not owned by the king were disafforested. More specifically free men could now graze their domestic animals 'within the forest at will' and could drive their pigs 'through royal demesne woods' to enable them to eat the acorns. The power of the ecclesiastical and lay lords to hunt was reaffirmed and 'every archbishop, bishop, earl or baron travelling through the forest may take one or two beasts by view of the foresters or he may blow his horn to give notice if they are not present.' Moreover, 'no man shall lose life or member for taking venison' and men outlawed for forest offences 'from the time of Henry II to the first coronation of Henry III' were pardoned. Although the Forest Charter reduced the power of the king, the struggle between royal power and that of landowners and local people who lived and worked in the areas under forest jurisdiction continued for many centuries.[35]

A medieval Royal Forest was a tract of land subject to the Forest Law, which was designed to protect the interests of the king, especially relating to hunting, timber trees and other rights. The standard legal definition was written by John Manwood, a barrister and forest official, in his *A Treatise and Discourse of the Lawes of the Forest*, first published in 1598: 'A Forrest is certen Territorie of wooddy grounds & fruitfull pastures, priuiledged for wild beasts and foules of Forrest, Chase and Warren, to rest and abide in, in the safe protection of the King, for his princely delight and pleasure.' The key point here is that no-one was allowed to hunt the wild animals protected by the forest law unless he had express permission from the king. The four beasts of the forest protected by the law were red, roe and fallow deer and wild boar. Areas of relatively wild land were maintained so that populations of these animals could be retained and within the forest, no matter who owned or held the land, 'only the king or those with his warrant were permitted to hunt these four beasts and there were restrictions on the use of the land and the woodland cover.'[36]

Royal Forests were often very extensive, and in England as a whole it is estimated that about a quarter of the whole country was under forest law in the early twelfth century. There were also many forests in Wales. But how was a forest organized and managed, and what did it mean to people living and working in the forest? We can examine this by considering one of the less well-known forests, the Royal Forest of Feckenham in Worcestershire. An important point is that the boundaries of forests could be changed. In 1086, for instance, large tracts of Worcestershire were within Royal Forest, which stretched from Herefordshire in the west across to Warwickshire. But by the thirteenth century 'two substantial areas had been disafforested' after the establishment of the Forest Charter in 1217: the Malvern Hills and surrounding land had become a chase or private forest in 1217, and large areas to the north and south of Worcester around Ombersley and Horewell had been disafforested by 1227.[37]

We know the boundaries of this and other forests because every so often a 'perambulation' around the boundaries was made to survey and fix them so as to ensure that everyone knew how far the Forest Law extended. The only surviving perambulation of Feckenham Forest took place on 30 May 1300 in the 28th year of the reign of Edward I, and this is particularly interesting as it describes how 'the smaller bounds' noted above were 'grudgingly accepted by the king' in a deal which 'was linked to the grant by parliament of a much-needed tax' of 'a fifteenth of all movables'.[38] In other words, the king had accepted a reduction in the forest area in exchange for a useful increase in taxation. The remaining area of the forest covered almost 200 square miles and included the royal manor of Feckenham and large parts of central and eastern Worcestershire, running up to the gates of the city of Worcester itself.

The main animals hunted from the thirteenth century onwards were the two larger species of deer: the red deer, which was native to Britain, and the fallow, which it is thought was introduced by the Normans, possibly via Italy.[39] The larger red deer were chased across open country and were ideal for the scale and terrain of Royal Forests, made up of large areas of open ground with patches of woodland (illus. 68). The smaller fallow deer were more normally kept as herds within deer parks owned and managed by the gentry. Fallow deer could be hunted with hounds, and the gentry could watch the hunt, but another method was for beaters to drive a group of deer towards a trap of waiting archers.[40] The wild boar had become 'relatively rare and localized in England, and they had probably disappeared from Feckenham Forest' by the twelfth century. In many ways the wild boar, largely immune to control by humans, was a better indicator of wildness than deer, but the distinction between the increasingly rare true wild boar and the domesticated pigs that thronged the woods becomes progressively difficult to pin down.[41] Henry III's request for twelve wild boars from the Forest of Dean in 1260 is perhaps the last reference to the species in England. However, there is a reference to a wild boar in the Plea Rolls of the Worcestershire Eyre of 1270 in which it was stated that 'a boar (aper) from the said forest followed some sows into the vill of Rous Lench' on 28 October 1266 and that two men from the village 'took the said boar and divided it between them'.[42]

The 'vert' of the forest, including trees, shrubs and grassland forming the wood pastures that were the habitat for the deer, was protected by law and those who lived in the forest had limited access to these resources. But they did have the right to cut wood and timber trees, and to graze their domestic flocks and herds. These rights, called estovers, were limited and strictly administered. Local people who held land could take wood for their own fuel, fencing or building repairs but not for sale to others. They could also use forest pastures for their own farm livestock, but the size of flocks and herds often caused disputes between local farmers and forest officials. Grazing was restricted when the female deer were fawning and pastures were closed off during what was known as fence month, centred on Midsummer Day, to allow the successful establishment of the new generation of deer. One of the most important restrictions was that people were not allowed to make new areas of arable and pasture land by clearing woodland, a process known as assarting, without permission. The growing population in the thirteenth century meant that there was great demand for new agricultural land and the king could manage this demand profitably by charging fines and rents on those who made new assarts.[43]

How were the royal forests administered and controlled? The main forest court was known as the eyre, and it was usually held in a local town or city. The officials of the forest eyre were justices who moved

19 George Stubbs, *Freeman, the Earl of Clarendon's Gamekeeper, with a Dying Doe and Hound*, 1800, oil on canvas.

20 *A Boar Hunt*, 17th century, circle/school of Peter Paul Rubens (1577–1640), pen and brown ink and brown wash.

21 South Indian painting of wild-boar hunting, *c.* 1775, gouache on paper.

from eyre to eyre dealing with different forests as they progressed around the country. An individual eyre could last several days and many local people were summoned to attend, including all lay and ecclesiastical holders of land in the forest: 'barons, knights and free tenants, together with four men and the reeve from every vill in the forest'. This was a cumbersome process and hundreds of people could be summoned. By the end of the thirteenth century, the gaps between eyres increased and at Feckenham, the eyre was replaced by more flexible 'inquisitions' concerning the state of the forest, eight of which were held between 1362 and 1377. The three main types of business heard at eyres and inquisitions concerned reports and sentencing of those poaching deer;

investigations of the extent of clearance of the forest by assarting; and reports of damage to trees.[44]

There was a complicated bureaucratic structure to document and report upon the various offences. Each Royal Forest had a keeper – at Feckenham the keeper was appointed, although in some forests the post was hereditary – and under this keeper were five hereditary foresters and a hereditary parker. These were wealthy and powerful figures who often appointed under-foresters to carry out administrative work for them, and the post could be very profitable. The foresters 'had many ways of exploiting their position' and some took 'bribes to permit breaches of the forest law such as the sale of wood and assarting' and exploited 'to their own advantage the very forest resources they were supposed to protect'. Another group of forest posts filled by members of the local gentry were those of verderer, regarder and agister. Verderers were elected court officers who worked with foresters to deal with poaching offences; regarders checked on the boundaries of the forest and inspected the condition of the forest; while agisters were particularly concerned with controlling the number of pigs in the forest.[45]

Where woods in the forest were owned by the king, local officials were appointed at the time of periodic sales of wood to supervise and keep records of the amount of timber felled and sold. Most woods in the forest were owned by private landowners who had to appoint an official woodward, whose role was to ensure that landowners did not exploit the wood and take more timber or wood than was customary. These woodwards were 'notable for being in the invidious position of being responsible to two masters' and when 'out in the forest on a daily basis' were 'subject to many temptations'. Moreover, if a woodward did not fulfil his role for the king the woodland could be 'confiscated and had to be redeemed by payment of a fine'. In 1280 it was reported that the two woodwards of William de Bello Campo at Alcester had damaged the woodland for which they were responsible by making excessive sales and gifts: 'those woodwards gave 1 cart of firewood for every 2 taken out.' The same year the parson of Grafton Flyford, with his brother and others, entered a wood at Hadzor 'at night time' and 'felled and carried away six oak saplings'.[46]

But sometimes individuals could disentangle themselves from the surveillance that so frequently trapped local residents. In the Plea Rolls of Worcestershire Forest Eyre of 1262, Peter de Lench Randulf went before the justices and argued that Robert Estrech the forester of Feckenham had 'impeded him in such a way that he cannot assart and improve to his own benefit a little grove (*gravetta*) next to his garden within his close in Rous Lench, which no forester ought to enter, as they say.' The verderers and regarders of the forest investigated and found that Peter's small grove was within his enclosed land in the village, which 'no forester or other ought nor is accustomed to enter except Peter.' The court ordered that no one in the future should obstruct Peter, 'who may assart the little grove and approve it for his own use whenever he wishes'.[47]

Most offenders were fined, or imprisoned and kept in prison until they had paid a fine. The records show that all types of people were likely to be poachers and although most brought before the eyre were men, there were several women. In 1264, when Margery de Cantalupo was staying at Studley Priory at midsummer in what was known as the time of grease, when male deer were considered at their fattest and best for hunting, her steward and others 'were accustomed to enter the forest with intent to offend against the king's venison. And they carried the venison they took to Margery, who knowingly received them.' Later in the same year Ralph, son of Constance of Coughton, 'took a buck in the said forest without a warrant, and carried it to Constance's house at Coughton, she knowingly receiving it'.[48]

The area to the north of Alcester and east of Studley was clearly prime poaching country. One of the most persistent offenders was Ralph Bagot, parson of the church of Morton Bagot, a still tiny village to the north of Alcester in Warwickshire. He was described as a 'common wrongdoer' and imprisoned and fined heavily in 1272, although he was soon released on bail under surety of his wealthy friends. This did little to stop his enthusiasm for hunting: he was accused of entering the forest on 9 September the next year 'with bows and arrows and dogs, to offend against the king's venison'. They, 'with others of their society, were common offenders' who 'were condemned time and time again'.[49] He was later identified as a member of a gang who had taken two bucks and carried them away on a cart on 30 December 1277, and was again imprisoned. But he did not stay there long as he was found guilty with several others, including Almaric le Despenser, lord of Oldberrow, of entering the forest on 18 October 1279 with a stalking horse, which they hid under and from which 'they shot two bucks and carried the venison wherever they wished.' All these records for the hunting parson were listed at a court meeting in 1280 and there is little reason to doubt that he continued to take the king's deer long after that date.[50]

The cumbersome forest eyre was replaced by inquisitions in the fourteenth century, but the lists of offenders continue. An inquisition of 29 April 1377 identifies various offenders, including William La Hunte of Astwood, who 'killed a buck in the place called "Arley" with a bow and arrow', while Thomas Jakettes with John Mauduyt killed a buck 'in the wood of the bishop of Worcester' and 'carried the flesh away with them'. Some people, such as Adam Salesbrugg of Hanbury, were multiple offenders. In April 1370 he 'took and had a doe on the night after Easter with various engines' and in 1371 he 'took and had a doe with his nets in the said forest'. Moreover on 11 November 1374 he took a pricket [a male deer in its second year] at night with his engines' and 'on various occasions, at night, entered the king's park of Feckenham with his nets and other engines to offend against the king's venison'. The view was that 'the said Adam is and for a long time was a common wrongdoer against the king's venison.'[51]

The administrative structure of the Royal Forests protected the interests of kings, who could derive useful income from the sale of rights to clear land for agriculture. But the importance of hunting for the maintenance of the power and authority of the king should not

22 Morton Bagot Church, Warwickshire, built in the late 13th century under Ralph Bagot, 2014.

be underestimated. Moreover the right to hunt in their private parks was an enormously important privilege and sign of status for the gentry. Medieval English kings 'built the great majority of their residences in rural settings, close to woods and wastes which formed the core of the royal forests' and whose main attraction was hunting.[52] Richard fitz Nigel, an important treasury official who later became bishop of London, noted in his *Dialogus de Scaccario* of 1176–7 that 'It is in the forests too that "King's chambers" are, and their chief delights. For they come there, laying aside their cares now and then, to hunt, as a rest and recreation.'[53]

In 1302 several of Edward I's staff went hunting in the castle park at Huntington in west Herefordshire on the Welsh border and wrote about the incident to the king. It seems to have been a 'carefully thought-out attempt' to use hunting as 'a theatrical vehicle for propaganda'. The king's agents were finalizing the 'formal take-over' of Humphrey de Bohun's estates and needed to 'obtain fealty of the tenants for their royal master'. This was an attempt 'to use a game reserve to reinforce the king's new control over a power-centre of one of his great tenants-in-chief, a place made more significant by its position on the edge of the Welsh March' where Edward I had been trying to impose his authority for ten years. A report written for the king by one of the royal hunters stated: 'because there is a fine (*beau*) park, we hunted barren does (*deymes baraignes*) therein the better to publish and solemnize your lordship (*seigneurie*) and seisin before the tenants and people of the country.' Here, it is clear, hunting was not merely a pleasure, but seen as a significant expression of the dominant power of Edward I in the face of crowds of people from the area.[54] The importance of royal power over hunting in Royal Forests was to some extent mirrored by the value that gentry placed on their ability to make parks for deer and to hunt in them. Such parks 'occupied a unique position in the social landscape as a meeting point for a variety of groups and a range of conflicting ideas'. As hunting was 'an activity closely associated with the assertion of social leadership and high standing', the enclosed parks 'aroused particular sensitivities in their contemporaries' because they enabled their owners 'to define who was to be involved with and, just as importantly, who was to be excluded' from the chase.[55]

Hunting could continue throughout the year, although different seasons favoured different styles of hunting and game. In the summer the male deer was particularly favoured as they were 'at their fattest and at their best for hunting, usually reckoned to be between Midsummer (24 June) and Holyrood Day (14 September)' – in the time of grease (*tempore pingwedino*) – and this was the period when royal hunting was most likely.[56] Female deer could be hunted in the closed season for stags and bucks, from around mid-September until early February, and other game species at different times of the year. King John and his queen, for example, 'spent much of March and April 1207 moving between various lodges and forests in central and south-western England, closely followed by the king's chief forest justice and master of hounds, Hugh de Neville, and his pack'. Over 150 years later, in January and February 1367, Edward III's winter itinerary was partially determined by visits to places that provided hunting.[57]

As we have seen at Feckenham, the 'great enthusiasm for venery spread far beyond a narrow court circle' and 'many of the nobility and gentry were keen hunters too, despite formal royal restrictions over hunting in many forest areas.' The poaching parson of Morton Bagot was participating in an activity highly favoured by the great churchmen such as Bishop Hugh du Puiset of Durham (1153–1195), an 'avid' hunter who made complicated arrangements with tenants to provide specific hunting services. Medieval account books show a telling diversity of hunting accoutrements such as bows, tents, hunting horns, dogs and hawks, and such items were frequently given as gifts. In 1304–5 'Henry de Lacy, earl of Lincoln, purchased a green tent and robe for his hunting trips.' Towards the end of the century Roger Mortimer, earl of March, purchased 'green hunting gowns for himself, his brother Thomas and a group of the king's chamber knights' and also bought 'a new bow, bow strings, and gilding for his freshly sharpened hunting knife'.[58]

This specialized equipment and clothing is persuasive evidence of the zest for hunting, which can indeed be seen as a form of spectacular conspicuous consumption. Enthusiasm was stoked by the need for monarchs, lords and knights to gain the skills of riding and using bows and arrows which for so long were central to warfare and were reinforced by ancient classical, biblical and contemporary ideas of chivalry which were constantly reworked in literature, ballad and song. The spectacle of hunting, its association with military prowess and horsemanship and the control of hunting grounds were key elements in the maintenance of princely, royal and aristocratic power and prestige. The historian Chris Wickham has argued that in the early medieval period the 'symbolism of hunting ever more clearly came to match that of royal charisma' and that the dispersal of forests and hunting rights to churches and aristocrats matched 'the dispersal of that charisma and of royal power in general to the aristocracy'.[59] In England and Wales under the Normans it was so important to medieval monarchs and landowners that they created a whole body of law to allow it to function and to provide a special landscape of forests and parks in which it might flourish.

THREE
Tree Movements

ENRY COMPTON, who has been described as 'something of a rootless cavalier', after a brief military career was bishop of London from 1675 until his death in 1713. He is most famous today for his role in the Glorious Revolution of 1688. He openly opposed James II's policy of employing Roman Catholic officers in the army and later signed the letter inviting William of Orange to England and also helped Princess Anne, later Queen Anne, to escape from London. The bishop of London's summer residence from the eleventh century until 1973 was Fulham Palace, which included a fine garden and parkland. Earlier bishops, such as Edmund Grindal (1558–70), who is thought to have planted the *Quercus ilex* which still survives (illus. 23), had already started to collect and plant trees and shrubs here. But Compton, who had 'a great Genius for Botanism', had the enthusiasm and opportunity to start collecting trees on a very extensive scale and 'apply'd himself to the Improvement of his Garden at Fulham, with new variety of Domestick and Exotick Plants'. In addition he was happy to share his knowledge and enthusiasm and showed 'great Civilities to, and had an Esteem for, all those who were anything curious in this sort of Study'. John Ray in his *Historia Plantarum* of 1686 listed fifteen rare trees and shrubs, many from America, in Compton's Fulham Palace garden (illus. 23), including the angelica tree and the tulip tree. Stephen Switzer considered him to be 'one of the first that encouraged the Importation, Raising and Increase of Exotics, in which he was the most curious Man of that Time'. Many years later, the Scottish botanist and garden designer John Claudius Loudon (1783–1843) described him as 'the great introducer of foreign trees' in the seventeenth century and stated that he 'may truly be said to have been the father of all that has been done since in this branch of rural improvement'.[1]

23 *Quercus ilex* (holm oak) at Fulham Palace, 2014.

But how did Compton manage to gather and grow so many new trees? The answer, at least partially, is that in his role as bishop of London, Compton was also head of the Church in the American colonies. He was conscientious in appointing and looking after clergy in the colonies and was able to appoint some who had a keen interest in botany. Switzer thought that 'by the recommendation of Chaplains into foreign Parts' he had 'greater opportunities of improving'. The most famous of these collectors was John Banister (1650–1692) from Twigworth in Glouces-tershire, who developed his interest in botany at Oxford, making use of the collections of the Oxford Physic Garden. He moved to Virginia in 1678 and almost immediately started to send specimens, drawings and lists of species to Henry Compton and other enthusiasts including John Ray and the gardener George London, who worked for Compton and was also a member of the Temple Coffee House Botany Club. By 1690 Banister had gained almost 1,800 acres of land in Charles City County, Virginia. He was a founder of the College of William and Mary in Williamsburg. But he was accidentally shot by a fellow explorer while plant collecting along the Roanoke River in May 1692. The new American species received at Fulham Palace included *Liquidambar styraciflua*,

Magnolia virginiana and *Acer negundo*. Compton himself died in his eighties and there 'were few days in the year, before the latter part of his life, but he was actually in his Garden' ordering the planting and care of his trees and plants.[2]

From the seventeenth century onwards the spread of tree species across the world gathered pace and many hundreds of tree species were introduced to Europe, especially from America and Asia. Trees became important commodities and new species had to be named, catalogued, tested and acclimatized. The great enthusiasm for novel trees led to the establishment of important tree collections, which from the early nineteenth century became termed 'arboreta'. Many of these, such as that of Earl Somers at Eastnor Castle, were developed by keen land-owners; others were established by botanical societies and by city councils, as at Derby, at least partly to educate the public.

Exotic enthusiasm

From the 1600s onwards the movement of tree species around the world and hence the variety of trees available to be planted in gardens, parks and plantations increased rapidly and dramatically. Four key stimuli to horticultural and silvicultural innovation in the period 1500 to 1900 have been identified. The first of these were scientific and technological advances such as the dissemination of botanical knowledge through publication of classical and modern works on plants and trees, experimentation, the work of botanical gardens, improvements in greenhouses and the development of the Wardian case, a sealed plant container which improved the survival rate of seedlings on lengthy voyages. Second, there were changes in attitude and taste and in particular the importance of fashions for particular tree species and styles of planting. Third was the development of an economic infrastructure: the successful establishment of important nurseries in suburban London such as Gordon at Mile End and Kennedy and Lee at Hammersmith and the establishment of tree nurseries on private estates, which assisted the rapid diffusion of newly introduced species. Finally there was the enormous increase in the number of introduced species. By 1550 it is estimated that there were 36 hardy and woody exotic species cultivated in England: 'by 1600, 103 species; by 1700, 239 species; by 1800, 733 species; and by 1900, 1911 species'.[3] Underpinning all these factors was the enormous growth in world trade and the associated development

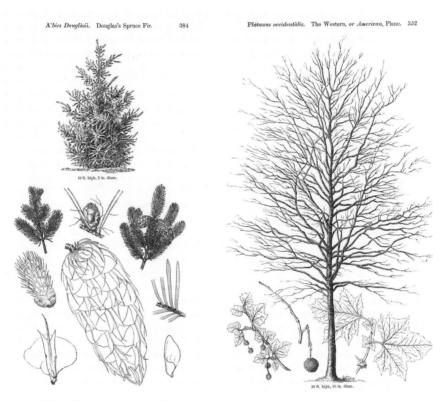

10 ft. high, 2 in. diam.

30 ft. high, 10 in. diam.

24 John Claudius Loudon, 'Abies Douglasii. Douglas's Spruce Fir', from *Arboretum et Fruticetum Britannicum; or, the Trees and Shrubs of Britain . . .* (1838).

25 John Claudius Loudon, '*Platanus occidentalis.* The Western, or American, Plane', from *Arboretum et Fruticetum Britannicum.*

of trading posts and colonies. European species were moved to Africa, America and Australia to make settlers and traders feel at home in their new surroundings, but far larger numbers of species were imported into Europe to be tested for their susceptibility to frost and their marketability as potential ornamental or timber trees.

British writers and horticulturalists were thrilled with the opportunities for profit and pleasure provided by these new trees. Mark Catesby, another collector to send plants to Henry Compton, extolled in his *Hortus Britanno-Americanus* (1763) the advantages of American trees that could be 'usefully employed to inrich and adorn our woods by their valuable timber and delightful shade; or to embellish and perfume our gardens with the elegance of their appearance and the fragrancy of their odours; in both which respects they greatly excel our home productions of the like kind'.[4] John Claudius Loudon went so far as to argue that 'no residence in the modern style can have a claim to be considered as laid out in good taste in which all the trees and shrubs are not either foreign ones, or improved varieties of indigenous ones.' He summarized the dates of introduction of foreign trees to the British Islands and the principal collections of trees in the second chapter of volume I of *Arboretum et Fruticetum Britannicum* of 1838.[5] This monumental work consisted of four volumes of dense text and four of drawings to illustrate and help identify all the different tree species currently known to be growing successfully in Britain. It both fixed the state of knowledge of individual trees and their movements across the globe at a particular time and was itself instrumental in encouraging further enthusiasm for collecting and planting new trees.

The widespread availability of popular translations of classical authorities in the eighteenth century also helped to encourage the planting and nurturing of trees. Virgil's *Georgics*, which had become very popular through John Dryden's translation of 1697, 'provided both a

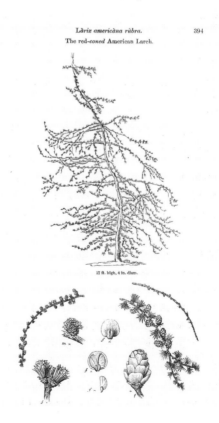

Làrix americàna rùbra.   394
The red-*coned* American Larch.

17 ft. high, 4 in. diam.

26 John Claudius Loudon, '*Larix Americana rubra*. Red-coned American larch', from *Arboretum et Fruticetum Britannicum*.

27 *John Evelyn Holding a Copy of 'Sylva'* (1687), 1818, engraving by Thomas Bragg after Sir Godfrey Kneller.

model for silviculture and an encouragement to the sort of botanical experimentation already taking place'.[6] The influence of John Evelyn's *Sylva* should not be underestimated, and although some later authors, such as Loudon, considered that he was 'more anxious to promote the planting of valuable indigenous trees, than to introduce foreign ones', successive editions of *Sylva* in 1664, 1670, 1679 and 1706 extolled the introduction of new trees. He 'had a voracious interest in new species, chiefly trees. Throughout his library any reference to the introduction of new species is marked or annotated.'[7]

Many enthusiastic botanists, such as Samuel Reynardson (who lived at Cedar House, Hillingdon, from 1678 until his death in 1721) and Dr Robert Uvedale (1642–1722) of Enfield, had large collections of exotic trees. Reynardson kept his trees mainly 'confined to pots and tubs, preserving them in green-houses in winter, never attempting to naturalize them to our climate'. They were not laid out in gardens and grounds in the form of planted arboreta.[8] The growing knowledge of introduced trees is shown by comparing Stephen Switzer's essays in *The Practical Husbandman* (1733) to his *Ichonographia Rustica* of 1718. In 1718 he recommended European trees such as oak, ash, beech, chestnut, hornbeam, Scotch pine, silver spruce, elm, lime and poplar. By 1733 Switzer was insisting that

> any one that would strive to bring the raising and planting of Forest Trees to their utmost Perfection . . . ought not to be content with treating barely on those plants that grow at Home, but ought by all means to endeavour at such Introduction of foreign Trees and Plants from Climates of equal Temperature, or (if possible) from Climates which are cooler than ours.[9]

The very rapid growth in the number of introduced species led to the practical need to identify, classify and label trees so that nurserymen, gardeners and owners could be relatively secure about which trees they bought and sold, discussed and displayed. New trees arrived in Britain first from Europe and Asia Minor, then in the seventeenth and eighteenth centuries from eastern North America, and finally a great surge from western North America, China, India and lastly Japan. Initially the classification and display of trees took place in a complex paper landscape of trade catalogues, botanical treatises and manuscript notebooks, descriptions taking the form of dried leaves and seeds, competing botanical nomenclatures and drawings of flowers, seeds, leaves and eventually whole trees. The innovative and influential binomial system of the Swedish botanist Carl Linnaeus (1707–1778) was central to these classificatory debates. Philip Miller (1691–1771), curator of Chelsea Physic Garden from 1722 until 1770 and author of *The Gardeners Dictionary*, which went through eight editions between 1732 and 1768, was for a long time reluctant to use the Linnaean system because he thought it would confuse gardeners, but he eventually used it in the influential eighth edition of his work, the last to be published in his lifetime.[10]

Loudon thought that in the seventeenth century the 'taste for foreign plants was confined to a few, and these not the richest persons in the community; but generally medical men, clergymen, persons holding small situations under government, or tradesmen'. In the following century, however, 'the taste for planting foreign trees extended itself among the wealthy landed proprietors' influenced by the Dowager Princess of Wales at Kew and by several aristocrats.[11] One of the most prominent landowners with a fascination for growing and displaying introduced trees in the eighteenth century was the Earl of Islay (later third Duke of Argyll) at Whitton on Hounslow Heath in Middlesex. The duke was best known 'as the personification of unionist Scotland in the first half century after union' but was also a keen classical scholar. He 'had one of the largest private libraries in western Europe' and was seen by some as one of the fathers of the Scottish Enlightenment.[12] One of Linnaeus's pupils, Pehr Kalm, noted that the duke was particularly interested in '*Dendrologie*' and visited Whitton in May 1747: 'there was a collection of all kinds of trees, which grow in different parts of the world, and can stand the climate of England out in the open air, summer and winter.' He pointed out that the duke had 'planted very many of these trees with his own hand', that 'there was here a very large number of Cedars of Lebanon' and that 'Of North American Pines, Firs, Cypresses, Thuyas' there was 'an abundance which throve very well'.[13]

Other key eighteenth-century arboreal enthusiasts include the ninth Lord Petre at his estate at Thorndon in Essex and as an advisor to the Duke of Norfolk at Worksop Manor, Nottinghamshire; the second Duke of Richmond at Goodwood, Sussex; Lord Bathurst at Cirencester Park, Gloucestershire; and the ninth Earl of Lincoln at Oatlands, Surrey. The last was described by the Duke of Richmond in a letter of 1747 as 'quite mad after planting'.[14] An early form of plant labelling is described by Dr Richard Pococke, who visited Lord Lincoln's Oatlands

28 Whitton Park, Middlesex, *c.* 1773, engraving from Robert Goadby, *A New Display of the Beauties of England: or, a Description of the most Elegant or Magnificent . . . Seats . . . in . . . the Kingdom* (1773).

on 29 April 1757. Visitors went down 'a winding walk' through shrubberies to a nursery 'laid out like an elegant parterre'. Near this was 'lately made another enclosure for all sorts of exotic plants that will thrive abroad, with boards plac'd over them on which their names are cut'.[15] Some of the plantings were on a very extensive scale. A letter of 1 September 1741 from the English botanist and gardener Peter Collinson (1694–1768) to the American botanist and explorer John Bartram (1699–1777) notes: 'The trees and shrubs raised from thy first seeds, are grown to great maturity. Last year Lord PETRE planted out about ten thousand Americans, which being at the same time mixed with about twenty thousand Europeans, and some Asians, make a very beautiful appearance.' Moreover 'great art and skill' was 'shown in consulting every one's particular growth, and the well blending the variety of greens'.[16] The nurseries at Thorndon were 'the most extensive private nurseries in the country' and after Lord Petre's death in 1742 at the early age of 29 the contents of his nurseries were sold to fellow tree enthusiasts: the Duke of Richmond, Lord Lincoln and Sir Hugh Smithson, later first duke of Northumberland.[17] From the late seventeenth century onwards well-known collections of trees on private estates were celebrated and much visited, but with the massive increase in tree species in the nineteenth century it became necessary for rapidly expanding collections to be formalized and displayed scientifically and artistically in planned, labelled and mapped arboreta.

Arboreta

The first clearly documented use of the term 'arboretum' to describe a collection of growing trees dates from 1796, in the original plan for a botanical garden at Glasnevin by the Dublin Society, which later became the Royal Botanic Gardens. The arboretum was established by Walter Wade, a medical practitioner who had published a flora of Dublin in 1794, and the head gardener, James Underwood. The garden was designed to be both didactic and practical. One area was classed the 'cattle garden'; it included distinct sections demonstrating different types of herbage that were 'injurious' or 'wholesome' for sheep, goats, 'horn cattle', horses and swine. The 'Hortus Linnaeus' was strictly educational and was divided into three sections: the herbarium, the fruticetum and the arboretum. The arboretum was established on the higher ground and took the form of a lengthy strip of trees along the southwestern edge of the botanical gardens. It was protected by a further 'plantation skreen' of trees on the outer edge of the gardens. The first catalogue lists more than 200 species of deciduous trees in addition to around 30 species of conifers.[18]

The regulations for planting and ordering the trees were strict. Each had to be 'arranged according to its Class, Order, Genus and Species'. Moreover every tree 'is to have a painted mark affixed to it', which indicated the number in the printed catalogue, 'the class and order' and the 'generic and specific name, all in black on a white ground, and the English name in red'. Great care was taken to identify species accurately and clearly in Latin and English. The arboretum was laid out to facilitate the viewing of individual trees. There was a 'broad gravel walk through the centre' of the arboretum and the grass was to be 'kept as fine as a bowling-green'. The specimen trees were 'to be planted from twenty to thirty feet apart, and where there is a very delicate or choice species, two may be planted, lest one should fail'. The uncertain nature of the requirements of newly introduced trees and the careful treatment they received is shown in the planting pattern chosen. The 'intermediate spaces' between the specimen trees were planted up with 'Fir, Larch, Laurel, Elm etc.' for shelter, but these plants were 'to be cut away when they come to interfere with the Linnean plants, or are useless as nurses'. Clear evidence of Wade's experience of practical plantation management and the danger when thinning trees of removing the best trees by accident is given by his stipulation that the 'nurses be as distinct in appearance as possible from the species they are planted to protect, as Deciduous for Evergreens, and *vice versa*.' So here, in the first place to be given that name, we can see that arboreta were from the start a combination of plantation, which usually consisted of a few varieties of trees, and botanical garden. Arboreta were Enlightenment projects that struggled to produce some sort of order and structure from the hundreds of new trees that circulated around the world and accumulated in collections. They were pedagogical sites that were usually linked with printed sources, such as guidebooks and catalogues, which informed visitors of the potential of the new species for enjoyment and commercial enterprise.[19]

Arboreta became especially popular from the 1830s as an ideal way to grow trees in larger private and public gardens, estate parks and botanical gardens. John Claudius Loudon is the person most strongly linked with the concept, but other gardeners and landscape gardeners were promoting the idea during the first decades of the century. George

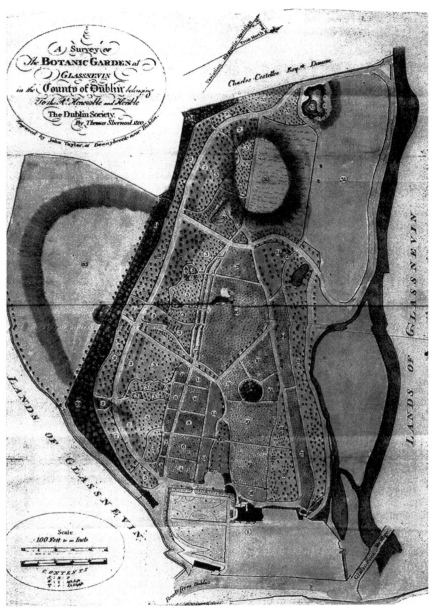

29 Thomas Sherrard, 'A Survey of the Botanic Garden at Glasnevin in the County of Dublin', 1800, from *Transactions of the Dublin Society* (1801).

Sinclair, head gardener at the Duke of Bedford's Woburn Abbey, under the auspices of the Society for the Diffusion of Useful Knowledge, enthusiastically recommended their widespread introduction. For Sinclair, the 'interest arising from the adoption of foreign trees into domestic scenery' was 'not confined to their picturesque effects'. Introduced trees reminded people 'of the climes whence they come' and the 'scenes with which they were associated'. In exploring 'a well-selected arboretum', he felt that the 'eternal snows of the Himalaya, the savannahs of the Missouri, the untrodden forests of Patagonia, the vallies of Lebanon, pass in review before us: we seem to wander in other climes, to converse with other nations'.[20]

Arboreta were likened to 'living museums' and were places where trials of new plants and experiments to discover the best growing conditions were undertaken. Arboretum staff sometimes aimed to replicate in microcosm the originating sites, and arboreta might be divided by taxonomy, climate, zone or geography. However, unlike laboratories, museums or glasshouse collections, arboreta were vulnerable to climatic and seasonal conditions and, in the context of rapidly industrializing Victorian society, the dire effects of air pollution. George Nicholson, curator of the Royal Botanic Gardens at Kew, produced during the 1880s a guide to the different kinds of trees and shrubs that could be planted in various conditions, including chalky, clay, sandy and peaty soils, marshy and boggy conditions and waterside. Tree collections in urban areas presented their own special problems and in his selection of trees and shrubs that could be 'best calculated to withstand the smoke and chemical impurities of atmosphere' within manufacturing towns, Nicholson tried to distinguish between those best adapted to withstand the industrial conditions of northern, Midland and southern towns. Across the vast expanses of the USA and Canada the variations were even more marked. Charles Sargent divided the North American continent from Arctic periphery to Mexican border into nine fundamentally different 'tree regions', defined according to the 'prevailing character of aborescent vegetation'.[21]

As the nineteenth century progressed, the movement of tree species around the world increased, with large numbers of new 'discoveries' made by European and American collectors. In the early twentieth century William Jackson Bean, curator of the Royal Botanic Gardens at Kew, when writing his comprehensive treatise on hardy British trees and shrubs, noted the 'enormous number of new species' that had become available for cultivation since Loudon's day through the activities of collectors such as William Lobb (1809–1864) in Chile and California and Robert Fortune (1812–1880) in Japan and China. Chinese varieties proved particularly challenging given that many remained unclassified or unnamed, while the designation of 'hardy' was difficult to apply as it depended upon gardening taste, economic value and experience. Neither were systems of nomenclature necessarily secure. By the early 1900s, particularly in Europe and the USA, attempts to subdivide species, genera and natural orders had become so prevalent that Bean feared it would 'involve such confusion and readjustment of nomenclature as to render its acceptance by cultivators' in Britain highly unlikely.[22] Disagreements concerning nomenclature were particularly evident where names celebrated political figures, nations or empires that were not universally popular, and there were 'numerous and perplexing' differences between 'European and American authorities' concerning the names of Coniferae.[23] These changes of names and classifications, in addition to the increase in the number of known species, made the role of arboreta in identifying, classifying and growing trees more important as well as more difficult.

Arboreta took many forms. Some, such as those at Eastnor Castle in Herefordshire, Elvaston Castle in Derbyshire and Westonbirt in Gloucestershire, were the products of the wealth, enthusiasm and hard work of successive generations of enthusiastic landowners and their gardeners and staff. Sometimes these followed precise and careful plans, but often they grew in an informal way, creeping over a large landed estate in the form of clumps of cedars, groves of American or Japanese trees and avenues of exotic pines. The rapid spread of new trees and shrubs into English arboreta is demonstrated at Eastnor Castle (illus. 30), where until the 1850s planting had been dominated by North and South American varieties, especially from California and New Mexico. These new introductions were tested by harsh winters and could fail. For example a hard winter in 1860–61 resulted in the destruction of 150 *Pinus insignis* and 130 *Cupressus macrocarpa*, which had reached heights of between 8 and 40 feet before they were killed by the frosts. It was soon found that some of the more tender trees that did not survive in the frost hollows at Eastnor could survive at higher elevations. After this very bad winter, the more tender species began to be replaced by specimens from Japan collected by John Gould Veitch and Robert Fortune, which seemed to cope with the harsh winters and the bad spring frosts to which, the gardener William Coleman noted, Eastnor was particularly prone. By the

30 Eastnor Castle
Arboretum,
Herefordshire,
2014.

later nineteenth century the arboretum became well known for holding
one of the best private collections of conifers in Britain. A visitor noted
in 1888 that the park scenery was 'of the greatest beauty and grandeur'
and that 'the wealth of conifers planted both as individuals and in groups
was almost unique in this country.' The owner of the estate, Earl Somers,
was particularly keen on cedars and had fine examples of the Japanese
cedar *Cryptomeria japonica*, 'beautiful thriving specimens of Cedar of
Lebanon' and a 'fine Mount Atlas Cedar' 50 feet in height which, 'like
many others here to be met with', had been raised from seeds gathered
by Earl Somers on Mount Atlas around 1859.[24]

Another type of arboretum was established by horticultural and
botanical societies which followed in the tradition of that founded
in Dublin in 1796. Local and national botanical, scientific and horticultural societies such as those at Glasgow, Edinburgh, Cork, Hull,
Liverpool, Manchester and Birmingham promoted botanical study
by forming arboreta that were available to members and the public on

various conditions. However, while these arboreta were intended for scientific societies, arguments often arose concerning the relative merits of aesthetics and botany. These were driven by the need to retain subscription income. The first English arboretum was established in the gardens of the Horticultural Society at Turnham Green in London from 1826. This was influential as it was visited by thousands of members, subscribers and their guests each year and also served as the most important centre for training gardeners and labourers. The arboretum was laid out by William Atkinson, who had assisted Sir Walter Scott in the landscaping of his gardens at Abbotsford in the Scottish Borders. Although established by a national scientific association with the important nurseryman George Loddiges sitting on the planning committee, the design of the Horticultural Society gardens was inspired more by aesthetics than taxonomy.[25]

The Horticultural Society was dependent upon the subscriptions of members and donations from patrons and therefore had to accommodate their wishes. Trees were arranged as 'clumps irregularly disposed upon turf' with further ornamental plants on the grass, while through the centre ran a canal supplied by a well in which aquatic plants were grown. The arrangement was admitted to be 'not systematic', although species of each genus were arranged 'as much as possible in the immediate vicinity

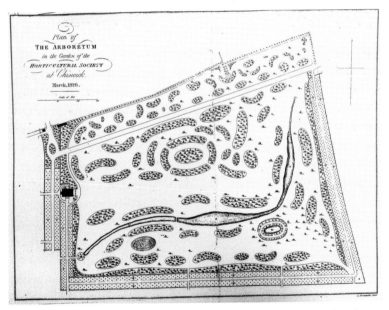

31 William Atkinson, *Plan of the Arboretum in the Garden of the Horticultural Society at Chiswick, March 1826* (1826).

of each other' to afford 'comparative examination'. The whole of the eastern sides consisted of separate groups of oak and elm plantations in front of a long strip that also included oaks and elms, while paths surrounded but did not traverse the arboretum.[26] The emphasis upon aesthetics rather than botanical principles was criticized by John Claudius Loudon, who noted that although the arboretum was the first of its kind in England, it was 'to be regretted' that the 'trees and shrubs were chiefly crowded together' in large, rectilinear clumps, which had 'subsequently never been sufficiently thinned out', rather than around a circumference of the buildings. As a result of this, 'the different kinds have not had an equal chance of displaying themselves, or of attaining that magnitude and character which they ought to have to answer the ends of an arboretum.'[27]

For Loudon the Horticultural Society garden was 'most defective in its general arrangement' and offered a 'want of grandeur and unity of effect as a whole, and of connection and convenience in the parts'.[28] He initially felt that the layout was 'so bad' that it could not be improved and should be 'totally obliterated'. But the trees were 'thriving' by 1833 and many of them had 'now achieved a considerable size', though this success caused a problem common to all tree collections, namely that the trees became too big for their situation. He noted that 'in most of the clumps they are crowding each other, so that the characteristic forms of individual species will soon be lost.' He thought that if instead they had been planted widely spaced the public could have gained useful knowledge of the 'forms, colours, and effect in the landscape' of different species. His critique was devastating: there was 'no thinking gardener' who could not foresee that soon 'this arboretum will become nearly useless for every purpose of the garden artist.'[29]

Loudon was able to put his own ideas into practice when he designed the Derby Arboretum, which opened in 1840 and was one of the first especially designed Victorian public parks. This arboretum was backed by the Strutts, the wealthiest Nonconformist manufacturing family in the region, and can be seen as a political counterpart to the aristocratic arboreta established nearby at Chatsworth and Elvaston. Unlike these, it was open freely to members of the public for two days a week. In encouraging the foundation of other public parks and arboreta, the Derby Arboretum helped to set the pattern for Victorian public urban parks, although it remained only a semi-public institution. The political, rational and recreational objectives of the Arboretum were emphasized

32 Nottingham Arboretum in a postcard of *c.* 1900. It was opened in 1852.

by Joseph Strutt in his speech at the opening ceremony in September 1840. It was a utilitarian venture, designed to be instructional, to improve the environment and to give pleasure to the citizens of Derby and elsewhere. Loudon specified that the tree species should be widely spaced and that once trees had become too big for their place, they should be removed. They were to be clearly labelled and many of the trees were placed on long, low mounds of earth to help create an illusion of size by obscuring walkers from each other on adjacent paths and helping to disguise boundaries. However, given that the mounds were not high enough to do this, their main function seems to have been to help demonstrate specimen forms and root systems.

The local perception of the value of Derby Arboretum was made clear by the Unitarian minister Noah Jones, who preached Joseph Strutt's funeral oration in 1844. The Arboretum, he said, was a 'noble gift' which would enhance the 'rational social pleasures of mankind', especially the 'toilworn artisan', encouraging him to work with greater diligence.[30] Health commissioners in 1845 considered that the park had 'already produced a perceptible effect in improving the appearance and demeanour of the working classes, and it has, doubtless, conferred an equal benefit upon their health'. It was argued that 'the most cursory' understanding of the life of the working classes would convince anyone of 'the immense advantages, moral and physical, that must accrue to the

inhabitants of closely-built towns by the establishment of public parks
. . . and gardens like that presented to the town of Derby'.[31] A *Derby
Mercury* editorial of 1851 thought that it was 'scarcely possible to over-
estimate the benefit which so large a space of ground for air and exercise
is capable of conferring upon a dense population'.[32]

Foreign visitors were also very impressed by this first public
arboretum. Charles Mason Hovey, who owned a nursery in Cambridge,
Massachusetts, and edited the *Magazine of Horticulture*, visited it in the
autumn of 1844 and thought it the best he had seen in England, Scotland
and France. He noted that there were no weeds, the flower garden was
flourishing and the tree labels were in full order. He considered that
Loudon's and Strutt's example should be followed, as 'we know of
no object so well deserving the attention of men of wealth than the
formation of public gardens free to all in crowded towns or cities.'[33] The
influential American landscape gardener Andrew Jackson Downing
thought it a 'noble bequest' that was 'in beautiful order' and 'evidently
much enjoyed' not only by locals but by strangers from the country
around. He met 'numbers of young people strolling about and enjoy-
ing the promenade', with nurses and children gaining strength from
the fresh air, while amateur botanists would carefully read the labels of
the various trees and shrubs and make notes in memorandum books.
Thus, in Downing's view, 'the most perfect novice in trees' could, by
walking around the Arboretum, obtain in a short time 'a very considerable knowledge of arboriculture while problems of identification and
classification could be solved by observation of living specimens'.
Shortly after his return to America in 1850, he set to work on a design
for extensive public grounds in Washington which incorporated many
of the ideas he had seen in Europe and included a garden of American
trees and a living 'museum' of evergreens. He also urged the creation
of a large park in New York. Downing considered that Derby Arboretum
was 'one of the most useful and instructive public gardens in the world'.[34]

The movement of trees, expertise, botanical theories and knowledge
between British, European and North American tree collections and
arboreta was complicated and reciprocal. America provided a major
source of exotic specimens planted on both sides of the Atlantic, and
American botanists utilized their knowledge of American natural history
to inform British botany and travelled between the two nations, maintaining contacts with scientific communities. American garden cemeteries
such as Mount Auburn became well known in Britain while American

landscape gardeners such as Downing and Frederick Law Olmsted travelled to Europe and made special studies of British institutions, horticulture and parks. British models for parks and arboreta were adopted that exploited the spaces and natural resources of the growing American empire. Rapid American economic growth, growing political and imperial power and urbanized population demands for leisure and education encouraged the formation of multiple national and urban parks. One result was the creation of more arboreta than any other country by the early twentieth century, many of which were associated with scientific and educational institutions, such as the Arnold Arboretum in Boston, Massachusetts. In Britain, by comparison, the idea of the arboretum became faded and old-fashioned and in 1961 Evelyn Waugh was able to identify it as pretentious:

> He faced, across half an acre of lawn, what the previous owners had called their 'arboretum'. Ludovic thought of it merely as 'the trees'. Some were deciduous and had now been stripped bare by the east wind that blew from the sea, leaving the

33 Thomas Chambers, *Mount Auburn Arboretum*, Massachusetts, mid-19th century, oil on canvas.

holm oaks, yews, and conifers in carefully contrived patterns, glaucous, golden and of a green so deep as to be almost black at that sunless noon . . .[35]

The Japanese larch in Britain

As areas of the world were opened up to trade in the eighteenth and nineteenth centuries, the range of new species of tree that could be tested for growth in Britain rapidly expanded. British landowners and foresters began to expect and demand novel trees that could be established on their estates and supplement the trees that they already grew. We can examine the cultural appropriation of a new species through the case of the Japanese larch. In the late nineteenth century the fashion for Japanese plants and gardens spread throughout Europe and the United States, part of a larger cultural moment that included the vogue for design, arts and crafts known as 'Japonisme'.[36] Japan was increasingly identified in Britain as a global partner, with a comparable imperial history and interest in horticulture. Japanese gardens were displayed in many international exhibitions and there was extensive coverage in the popular press, including the numerous illustrated garden magazines. This led to a flourishing export market in mature plants, bulbs and stone ornaments. In Japan large export nurseries were established, such as L. Boehmer's and the Yokohama Nursery Company. British-based nurseries stocked an increasing range of Japanese plants and ornaments, as did the leading Arts and Crafts design store Liberty's of London, which advertised Japanese stone lanterns.[37]

Before 1858, when Japan opened three treaty ports to the West, it had been largely secluded from the world from the early seventeenth century, except through links with the Netherlands and China via the artificial island of Dejima at Nagasaki. Most plant introductions and knowledge in Europe about Japanese plants arose through Dutch trading connections. The botanical writing and descriptions of Engelbert Kaempfer (1712, 1727), Carl Peter Thunberg (1784) and Philip Franz von Siebold (1850) played a major role in raising expectations about the new plants that could be found once trade took place more regularly.[38] The European collectors derived much of their knowledge from collecting plants, botanizing with local experts and consulting texts, illustrations and herbaria written, drawn and collected by them.[39] Early Victorians therefore had some tantalizing knowledge about the diverse flora of

Japan and also had considerable experience from earlier introductions of the likely potential of Japanese plants in British gardens and parks. It was, however, only after 1858, when Japan opened three treaty ports to the West, that British collectors and horticultural traders were able to experience the Japanese landscape at first hand and exploit fully the commercial potential of its trees and shrubs.

Western names were applied to plants that already had established Japanese botanical and horticultural names, although these latter were also used by Western nurseries to reinforce the exoticism of newly introduced Japanese plants. The Japanese name for *Larix leptolepis* is frequently given as *karamatsu*,[40] although in the important Japanese plant encyclopaedia *Kai* (1763) this larch is given three different names: *Kin sen shou*, *Fuji matsu* and *Nikko matsu*. The first name refers to the golden colour of the autumn foliage, while 'Fuji' and 'Nikko' are both areas where the tree grows naturally and *matsu* means pine. The tree is described as having a thick, scaly bark similar to the Japanese white pine. It is noted that after frost the needles fall, and the golden autumn colour of these needles is particularly praised (illus. 34).[41] The tree had a limited distribution, mainly at altitudes of between 1,000 and 1,400 metres, especially on dry, volcanic soils. The main natural stands were found in central Honshu, particularly in Yamanashi Prefecture, including the slopes of Mount Fuji. But larch had long been planted in northern Honshu and many more plantations were made in southern and central Hokkaido after colonization.[42] The Japanese larch, though known to Kaempfer and Thunberg in the eighteenth century and mentioned by Lambert, was first described by Lindley in 1833. The tree was included in Siebold and Zuccarini's *Flora Japonica* (1843) and illustrated with a spray of twigs and needles, and cones with details of the needles and seeds, but it was not actually grown in Britain until the 1860s.[43]

John Gould Veitch (1839–1870), a leading British plant collector, arrived in Japan in 1860. He worked for his family company at the Royal and Exotic Nurseries based in Exeter and Chelsea. The *Gardeners' Chronicle* of 15 December 1860 reported that 'Japan stands pre-eminent' as a place where the 'vegetation is vigorous, and little known'. It had a 'climate like that of England, and a half Siberian or Himalayan and half Chinese Flora' and 'offers the greatest inducement to Europeans to investigate its productions'.[44] In September 1860 Veitch discovered what he called *Abies leptolepis* with three other conifers, and his climbing companion, the diplomat Rutherford Alcock, reported that

34 'Kin sen shou', illustration of 1763 from Shimada Mitsufusa and Ono Ranzan, *Ka-i* (1759–63).

it grew 'at an elevation of 8,000 to 8,500 feet on Mount Fusiyama'. He thought it was 'remarkable as being the tree which grows at the greatest elevation on this mountain'. They found some specimens 40 feet high but 'on ascending the mountain' it 'dwindles down to a bush of 3 feet'. He noted that the 'Japanese name is *Fusi matsu*'.[45] Initially there was confusion as to whether the Japanese larch discovered by Veitch was the same species as the *Larix leptolepis* described by earlier botanists. The *Gardeners' Chronicle* of 12 January 1861 noted that the cones described in Siebold and Zuccarini were 'four times larger than those sent home by Mr Veitch' and that 'there is some doubt whether his plant is not distinct'. Elwes and Henry stated that 'A stunted form, growing on the higher parts of Fuji-yama, was collected by John Gould Veitch, and was considered to be a new species by A. Murray; and is recognised as a variety by Sargent.'[46] They provided eight different botanical names for the Japanese larch,[47] eventually plumping for *L. leptolepis*, as 'it is the name by which this species is universally known'. They argued that the use of *L. kaempferi*, as proposed by Sargent, 'would cause great confusion, as this has been used for *Pseudo-larix Kaempferi*, the golden larch of China'.[48]

The difficulties over naming the newly imported tree did not hinder its appreciation by foresters and landowners. Although few trees were established from the seeds collected by Veitch in 1861, trees grown from other seeds 'grew so well generally that it is now being planted almost everywhere, and some of the older trees have produced good seed for ten years or more'. Elwes was keen to assess the growing conditions and uses of the tree in Japan and its suitability as a tree for forestry plantations. He saw the trees growing in volcanic soils in Japan in 1904 and thought they 'were very similar in habitat to the larch in the Alps, and had not an excessive development of branches'. He noted that the timber was used for 'ship- and boat-building' and 'railway sleepers and telegraph poles'. The plantations in Japan were also closely connected to the demands of modern development. Elwes saw many young plantations which 'were very similar to larch plantations in England in growth and habit. I also saw it planted experimentally in Hokkaido, along the lines of railway, where it seemed to grow as well in this rich black soil as in its native mountains.'[49]

The tree became very popular in Britain and was 'looked upon by many foresters as likely to replace the common larch'. No recently introduced conifer had 'attracted so much attention among foresters as the Japanese larch, which, during the last ten years, has been sown very largely by nurserymen'. Elwes himself successfully sowed seeds collected from trees from three different British estates in 1890 and after six years they had grown to 4–8 feet in height. In his view the Japanese larch had three main advantages. First, its establishment as a plantation at 1,250 feet in Scotland, where it grew 'very vigorously in mixture with Douglas fir', showed it to be hardy. Second, it appeared to be immune from the canker *Peziza willkommii*, which affected European larch. Henry examined in 1904 'six plantations of Japanese larch of ages from five to sixteen years, and in none could detect any sign of canker'. Third, it was a vigorous tree suitable for economic plantations, as it grew in its first twenty years quicker than European larch, although it appeared to have 'a great tendency to form spreading branches'.[50] By the mid-twentieth century Japanese larch had become 'one of the most important exotics planted in Britain' with about 14 million seedlings planted annually, a number exceeded only by Sitka spruce and Scots pine.[51]

The movement of Japanese larch to Britain demonstrates how there were often several stages in the reception and transculturation of a new tree as it was named, collected, transported, acclimatized and naturalized. The stage when new trees were conceptualized as exotic can itself be recognized as one form of enculturation. A second stage is when the tree became 'culturally assimilated'. At this stage the trees are not physically changed or modified, but they have been grown long enough to demonstrate that they are well adapted to live in the open air, and propagate well enough to become common plants in Britain. A key factor in allowing a plant to become culturally assimilated is its hardiness. Hardy plants were useful for ornamental and commercial planting in existing gardens, parks and woods; they became common in the British landscape and were no longer seen as exotic. In some cases there was a third stage of physical hybridization. The crossing of exotic and native species of plants was one way to produce new varieties of plants; sometimes, however, hybridization occurred naturally. *Larix leptolepis* became culturally assimilated through its economic timber value and eventually became physically hybridized with the European larch at Dunkeld in Scotland to produce *L. eurolepis*, which itself became an important commercial tree species.

In 2003 the National Inventory of Woodland and Trees showed that the three larch species together formed about 10 per cent of coniferous woodland in Britain and 5 per cent of all woodland. The same year the Forestry Commission reported on a breeding scheme to improve the genetic quality of larch and noted that both European and Japanese larch were 'popular with foresters' and had gained 'a reputation as a good nurse for hardwood species; casting a light shade so allowing ground vegetation to persist; a degree of fire tolerance and high amenity value'. The timber was 'valued for good durability' and was used 'in the manufacture of outdoor furniture, bridges and fences as well as boatbuilding'. This report noted that the 'Japanese variety was observed to generally grow faster than its European counterpart, although with perhaps poorer stem form, and showed little sign of shoot canker which can be a problem' for European larch.

This promising position, however, was very soon to change, for in February 2002 five container-grown plants of the well-known garden shrub *Viburnum tinus* 'Eve Price' from a garden centre in West Sussex were spotted by Defra Plant Health and Seeds Inspectors to be suffering from 'severe aerial dieback, stem base discoloration and partial root decay'. One plant was sent to be tested and was found to be infected by the fungus-like pathogen *Phytophthora ramorum*, which was formerly unknown in Britain. It is not known where the pathogen originated, though there is some evidence that it is native to Asia. Once the pathogen was known to exist in Britain it was searched for and found to be infecting a wide range of exotic garden shrubs, including rhododendrons, camellias and viburnums, but in 2008 it was also found to be infecting the bilberry *Vaccinium myrtillus*, a native plant of considerable conservation importance and characteristic of moorland and heathland. Until 2009 very few trees had been affected by the disease, and these were mainly ornamental conifers growing in gardens or trees growing close to infected stands of the common *Rhododendron ponticum*, which was in the past widely planted in woodland as cover for game. But in August 2009 many thousands of Japanese larch trees in southwest England were discovered to be infected by the disease. The Forestry Commission note that this was 'the first time it had been found causing lethal infection (in the form of stem cankers) on a commercially important conifer species anywhere in the world'. The disease was later confirmed in European larch (*L. decidua*) in Cornwall in March 2011 and has also been found in young Douglas fir of about five to ten years of age.[52]

Suddenly, the whole future of Japanese larch, which had developed into a welcome and secure additional British timber tree, was thrown into doubt. The name 'Phytophthora' comes from the Greek for 'plant destroyer' and it is one of the most destructive genera of plant pathogens. It was soon discovered that 'Japanese larch can produce very high quantities of disease-carrying spores when actively growing in spring and summer, at much higher levels than those produced by rhododendron.' *Phytophthora ramorum* can spread rapidly over long distances and the wet and damp atmosphere of western Britain encourages this spread. The Forestry Commission notes that the disease can be spread 'by animals and possibly also by birds, as well as on footwear, vehicle wheels, and machinery and equipment used in forests'. In addition to infecting larch needles, the disease infects the tree's bark, resulting in dieback and the death of the tree. This takes place very rapidly and the disease is known to kill trees 'within one growing season after its presence is first detectable'. The main symptoms of the disease are the wilting of the tips of the shoots, premature shedding of the blackened needles and the development of bleeding cankers on the upper trunk and branches. As no cure has been found, the only treatment is for owners to cut down all infected larch trees as quickly as possible. Many thousands of acres of larch have been felled.[53]

The movement of trees around the world has brought enormous benefits. Gardens and parks are more diverse and varied and we can touch, smell, study and enjoy trees from across the globe. The range of trees available for commercial timber production has been greatly enhanced. But there are also some negative and even dangerous consequences. The excitement about the beauty of exotic trees can soon turn to hackneyed acceptance. Popular garden trees in Europe such as the American false acacia (*Robinia pseudoacacia*) can invade the countryside, damage valuable conservation woodlands and prove almost impossible to eradicate. Extensive monocultures of introduced eucalyptus and spruce may irreversibly destroy important wildlife habitats. Perhaps most dangerous, however, is the modern way of moving nursery stock around the world. Foreign trees used to be grown from seeds and cuttings; nowadays they are more likely to be moved around from country to country as growing individuals in containers. This allows diseases such as *Phytophthora ramorum* and *Chalara fraxinea*, which causes ash dieback, to spread rapidly with disastrous consequences.

FOUR

Tree Aesthetics

THE more adventurous British travellers to Italy started to visit Sicily in the eighteenth century and one of the first and most influential British guides to the island was Patrick Brydone's tour of Sicily, published in 1773. This was based on a visit he made in 1770 and it aroused controversy, since his interpretation of the local priest and eminent naturalist Giuseppe Recupero's observations on the layers of lava suggested that the lowest strata must be at least 14,000 years old, considerably older than what was then the accepted age of the Earth. Dr Johnson thought that if 'Brydone were more attentive to his Bible, he would be a good traveller'. Brydone was also an observant commentator on society and natural history. When climbing Etna, following breakfast at Piedmonte, they rode on mules through 'some beautiful woods of cork and ever-green oak, growing absolutely out of the lava, the soil having as yet hardly filled the crevices of that porous substance', and eventually reached 'the great chesnut-trees', many 'of an enormous size; but the Castagno de Cento Cavalli is by much the most celebrated'. Brydone had found it 'marked in an old map of Sicily, published near an hundred years ago; and in all the maps of Ætna, and its environs, it makes a very conspicuous figure'. But he was disappointed and 'by no means struck with its appearance, as it does not seem to be one tree, but a bush of five large trees growing together'.

Brydone complained to his guides that this was not a true single tree, but they 'unanimously assured us, that by the universal tradition and even testimony of the country, all these were once united in one stem'. The British party then 'began to examine it with more attention' and found that 'there was indeed an appearance as if these five trees had really been united in one' as there was 'no appearance of bark on the inside of any of the stumps'. They measured it to be '204 feet round' and

thought that 'if this was once united in one solid stem', it must have
been 'looked upon as a very wonderful phænomenon in the vegetable
world.' Moreover, he was subsequently told by the 'ingenious ecclesi-
astic of this place' Giuseppe Recupero that when he had paid for
'peasants with tools to dig round the Castagno de Cento Cavalli' he
found that 'all these stems united below ground in one root.' Brydone
noted several other huge chestnut trees nearby and thought that the
size of the trees was encouraged by the 'vast quantity of nitre' in the
'thick rich soil' formed from 'ashes thrown out by the mountain' and
'the smoke of the volcano', which 'must create a constant supply of this
salt, termed by some, not without reason, the food of vegetables'. That
the tree was once an important supplier of chestnuts was indicated by
the 'ruins of a house inside of the great chestnut-tree which had been
built for holding the fruit it bears', where the party 'dined with excellent
appetite'.[1]

Seven years after Brydone's visit the young, wealthy connoisseur
and tree enthusiast Richard Payne Knight (1751–1824) left Rome on
an expedition to Sicily with the German landscape painter Jakob Philipp
Hackert and the wealthy amateur painter Charles Gore, who later lived
at Weimar and became a great friend of Goethe.[2] Knight was on his
second visit to Italy, while his extraordinary new gothic castle at Downton
(1772–8) was being built near Ludlow, and travelling with the landscape artist John Robert Cozens. The aim of the trip was to record
and measure the Greek temples, but they also climbed to the crater of
Etna. Knight intended to publish his account of the journey with illustrations, but this was not completed, and it was not until 1980 that
Claudia Stumpf rediscovered the original English text at Weimar. On
1 June 1777 Knight's party went a few miles out of the way to 'see the
famous Chestnut-trees in which we were much disappointed'. The tree
called '*la Castagna di cento Cavalli*' was 'not a single tree but a groupe,
and the rest, tho' very large, are all Pollards very low and much mutilated'. Knight noted acidly that in 'Sicily they might be looked upon
as wonders, as a great part of the Inhabitants never saw a tree larger than
a Dwarf Olive' but to people like him 'who have been used to the
noble Oaks of England, they are very contemptible objects'.[3]

Unfortunately there is no surviving drawing by Knight's travelling
companions Gore or Hackert of this tree, but Jean-Pierre Houël made
one at roughly the same time which shows the chestnut barn within
the tree.[4] What is clear from the descriptions of the two British visitors

is that their aesthetic ideal of a worthy tree was one that had a large single stem and which had considerable height. Brydone was clearly relieved when he received the evidence from his friend Recupero that the peasants' excavations indicated that the tree had a single root, but his initial lack of enthusiasm upon coming upon this clump of five trees rather than one single stem remains dominant. In addition, the accounts show that the ideal tree should not be tampered with by human activity, should not be 'mutilated' by pollarding or cropping; such trees had no right to be called 'wonders' but were merely 'contemptible objects'.

But how can this contempt for the mutilated pollard be squared with the enthusiasm shown by painters such as Jacob van Ruisdael and Thomas Gainsborough, whose representations of such trees gained great popularity in the seventeenth and eighteenth centuries? To understand this apparent anomaly we need to consider the relationship between art and trees and the influence of specific artists on the aesthetic interpretations of trees. This relationship provides a context for the popularity of the Picturesque in the eighteenth century. Authors such as William Gilpin and Uvedale Price encouraged the celebration and enjoyment of ancient, crooked trees in remnant areas of wood pasture and common and old pollarded willows growing along rivers. Price was particularly influential in encouraging delight in old, rugged, even decaying trees. He was concerned that new plantings and woods, including introduced species, should fit in with the existing landscape, and his ideas were taken up by William Wordsworth, Walter Scott and many other influential nineteenth-century figures.

Drawing and painting trees

Early representations of trees in paintings are generalized, with little indication of the species of tree concerned. The form of the tree is, however, quite frequently indicated and it is possible to distinguish fairly clearly whether a tree has been pollarded. One of the earliest representations of cut trees is from a thirteenth-century Latin text on the Welsh Laws which shows both coppiced and shredded trees.[5] The series of paintings in the illuminated manuscript *Très riches heures*, made for Jean, Duc de Berry, in 1411–16 are some of the earliest true-to-life landscape paintings showing trees. Most of the trees, whether pollards growing along a river or next to a pool, trees in a wood with a man wielding an axe next

35 Limbourg brothers, 'December', from the *Très riches heures* of the Duc de Berry, 1411–16.

to a pile of faggots or dense trees surrounding a woodland clearing used for hunting, indicate careful management. In most cases, however, the tree species are not identifiable, although the pollards adjoining water are most likely willows and the greyish bark and dense covering of brown leaves on the floor of the wood in the hunting scene are reminiscent of a beech wood. The wonderful image of pigs eating acorns on the edge of a dense wood shows that the trees must be oaks, although they would be unidentifiable were it not for the carefully depicted acorns.[6]

Although trees are important features of many Renaissance paintings and drawings, most are not identifiable as individual species. One of the earliest artists to draw recognizable tree species was Albrecht Dürer (1471–1528), who made several drawings from nature of trees near to his home in Nuremberg. In *c.* 1497 he painted one of his finest watercolours, a study of a single spruce tree (*Picea abies*), which is thought to have been made in the 1490s on his return home from northern Italy, where he had been influenced by such artists as Mantegna and Bellini (illus. 69). The tree is painted as an individual subject with no background landscape, and Dürer brilliantly captures the colour and form of an individual spruce. This is an example of one of his drawings from nature, which he used to help design landscape backgrounds for his various paintings and woodcuts. Another watercolour of a similar date is the enigmatic study of water, sky and pine trees characteristic of the landscape near Nuremburg. This again clearly depicts a specific tree species, the pine (*Pinus sylvestris*), with its dark blue-green foliage and reddish bark, but here the trees are in a group and set within a landscape context of sandy soil and a pool. The watercolour is probably unfinished and the odd group of truncated trees adds a sense of desolation to this atmospheric portrayal of trees in a landscape. It is likely that some of Dürer's work had an influence on the German printmaker and painter Albrecht Altdorfer (1482/5–1538), who also depicted identifiable trees in his novel and experimental landscape etchings of around 1520–21. It has been argued that his 'success in rendering a sense of airy atmosphere in

36 Albrecht Dürer, *Pine Trees*, *c.* 1497, watercolour and bodycolour.

37 Albrecht Altdorfer, *Landscape with Spruce and Two Willows*, 1520–21, etching.

his landscape etchings is quite unlike anything to be seen in Dürer's work'
but his depiction of trees is less assured. In his *Landscape with Spruce and
Two Willows* the tree species are identifiable more from the form of
their trunks than their foliage: the two willows have both been recently
pollarded and have fresh regrowth, and their lush and artificial form
contrasts with the central trunk of the spruce.[7] One of the most remarkably realistic scenes of a woodland interior is by the Flemish artist Pieter
Bruegel the Elder and is dated to the mid-sixteenth century (1540– 69).
This shows a family of seven bears around what appears to be their lair
in the roots of a tree. The woodland consists of a mixture of standard
deciduous trees with several coppiced trees in the middle distance. The
range of tree forms is splendidly captured, as are details such as the
single live stem growing from the otherwise decaying and hollow tree
to the left; but the tree species cannot be guessed.

Art historians argue that one of the greatest achievements of Dutch
artists of the seventeenth century was making landscapes acceptable
as serious art by patrons, collectors and the cognoscenti. Although
Rembrandt is not best known for his landscapes, several of his drawings,
paintings and etchings contain evocative representations of trees. His
St Jerome beside a Pollard Willow of 1648 was entertainingly described

38 Pieter Bruegel the Elder, *Woodland Scene with Bears*, 1540–69, pen and brown ink over black chalk.

in 1932 as a 'tree study with Saint Jerome thrown in'. The tree certainly appears to be drawn carefully from nature, with the bark growing back over the damaged trunk and the young branches sprouting from cut boughs. Marijn Schapelhouman questions this and argues that one 'certainly cannot rule out the possibility that Rembrandt simply "made up" his pollarded willow', and also emphasizes how St Jerome 'is part of a coherent triangular composition' although the 'melancholy lion [that] looks out at the viewer from behind the tree with a somewhat bewildered expression' does not fit into this scheme. Dead or dying trees

39 Rembrandt, *St Jerome beside a Pollard Willow*, 1648, etching and drypoint.

40 Rembrandt, *The Three Trees*, 1643, etching with drypoint and burin.

could represent the transience of life and for some Christians could be interpreted as the Tree of Life in the Garden of Eden, which after the Fall became the Tree of Death. St Jerome noted that Christ's cross was made from the wood of this tree, which symbolized redemption through faith. Moreover the freshly sprouting leaves of the pollarded willow, which is not actually of course dead, could be interpreted as a symbol of the Resurrection.[8]

Perhaps Rembrandt's most famous representation of trees is *The Three Trees* of 1643, which shows a carefully delineated landscape populated with farm workers and travellers and a distant town under a stormy sky. Concealed and sheltered in the vegetation below the three trees are two lovers; the trees are starkly set against the light sky. The trees are very difficult to identify: different botanical experts have recognized them as willows, elms, birches, oaks and beeches. They have often been associated with the three crosses at the crucifixion of Christ. Cynthia Schneider argues that this etching influenced Jan Lievens's painting *Landscape with Three Trees* of *c.* 1645. This may be true, but the form of the dark and gnarled three trees painted by Lievens is very different; they appear to be ancient overgrown oak pollards that have been uncut for many years, which provide dense shade for those resting underneath.[9]

Jacob van Ruisdael (1628–1682) has been described as 'the father of tree illustration'. Although many of the trees in the backgrounds of his landscapes are unidentifiable, individual trees and groups of trees in the foreground are usually clearly recognizable as oaks, elms, beech and willow. A team of botanists and art historians at Harvard searched many works of the Dutch school but 'failed to unearth a single master before Ruisdael who depicted a variety of trees using a suite of independent characters that are botanically diagnostic'. Ruisdael was born in Haarlem and lived there until around 1656 when he moved to Amsterdam. The identification of his trees is helped by the fact that most of drawings and paintings were made close to his home, where the sandy soils supported a fairly limited range of trees. Moreover, this was before the major period of plant introductions and so most of the trees depicted are those that had traditionally grown in the area. The complete catalogue of Ruisdael's paintings, drawings and etchings demonstrates that there are over 150 works of which woods and forest are the main theme. Trees and groups of trees are also vital features in his paintings of many other subjects, such as mills, cornfields and waterfalls. In his early period Ruisdael concentrated on the countryside near Haarlem, with its 'crowded woods with sandy footpaths and tangled

41 Jan Lievens, *Landscape with Three Trees*, c. 1645, oil on panel.

42 Jacob van Ruisdael, *Landscape with a Cottage and Trees*, 1646, oil on panel.

trees set in thickets'. His later forest pictures are more spacious, with more clearings, openings and distant views. Several of his paintings include old fissured tree trunks and decaying branches, and these have been interpreted as symbols of fragility and corruptibility, although the art historian Seymour Slive questions whether Ruisdael 'intended them to have one of the multitudinous meanings that have been assigned to them since the tree of life and tree of knowledge appeared in the garden of Eden'.[10]

One of his earliest paintings, *Landscape with a Cottage and Trees*, dated 1646 and painted when he was around eighteen, shows a dense group of trees in the right foreground obscuring a small cottage. Two trees dominate: an old and decrepit pollarded willow next to a small pond, and an oak. The pollard is very accurately painted and clearly shows the stubs where several branches have been cut; the foliage and leaves of the oak are also very carefully depicted, and the trunk has characteristic orange and white lichen. To the left of the cottage are two young, healthy and vigorous willow trees. A few years later Ruisdael made several etchings, of which *The Three Oaks* (1649, illus. 43) demonstrates his astonishing ability to capture the form of the trunks, branches and foliage of the oak tree. A similar group of oaks appears in his painting *Wooded Landscape with a Pond*, which was in the Picturesque

43 Jacob van Ruisdael, *The Three Oaks*, 1649, etching.

enthusiast Richard Payne Knight's collection at Downton Castle, Herefordshire. Another etching of a forest scene dated about 1650–55, *Forest Marsh with Travellers on a Bank*, includes a depiction of a remarkable oak with several branches bare of leaves. This is an accurate representation of what happens to established and old oak trees where the water table has risen. This etching was influential with succeeding artists: Ruisdael's pupil Meindert Hobbema painted a version of it in 1662, and well over 100 years later the young John Constable wrote to his tutor John Thomas 'Antiquity' Smith that 'I have a great mind to copy one of Rysdael's etchings, I have seen one at your house where there are two trees standing in water.'[11]

Thomas Gainsborough was an enthusiast for drawing trees and landscapes from an early age. He recalled that

during his Boy-hood, though he had no idea of becoming a Painter then, yet there was not a Picturesque clump of Trees, nor even a single Tree of beauty . . . for some miles around the place of his nativity, that he had not so perfectly in his mind's eye, that had he known he could use a pencil, he could have perfectly delineated.

The first drawing he remembered making was of a group of trees, and in one letter Gainsborough went so far as to argue – probably, as the art historian Michael Rosenthal points out, to tease the recipient – that historical figures in a landscape painting could be placed 'for the Eye to be drawn from the Trees in order to return to them with more glee'. Trees are not only important in Gainsborough's landscapes but also very frequently occur as backgrounds to portraits and peasant scenes. At the end of his life he felt a great 'fondness for my first imitations of little Dutch Landskips', and his copy in chalk made *c.* 1747 of Ruisdael's *La Forêt* or *Wooded Landscape with a Flooded Road* survives. No print of Ruisdael's painting is known to have been made, and it is possible that Gainsborough copied the original.[12]

Gainsborough's trees had an enormous impact on the development of Picturesque aesthetics. Uvedale Price's grandfather Uvedale Tomkyns Price was one of Gainsborough's most important patrons at Bath; their friendship developed around 1758 when Price was in his seventies and Gainsborough was 31.[13] Price's portrait by Gainsborough (1761–3) shows him seated next to a collection of drawings and holding a drawing of trees with cut branches with his left hand and a porte-crayon with his right. Gainsborough visited Price's Herefordshire estate, Foxley, in 1760 and his *Beech Trees at Foxley* is signed and dated in that year. This

44 Jacob van Ruisdael, *Forest Marsh with Travellers on a Bank,* 1650–55, etching.

45 Thomas Gainsborough, *Beech Trees at Foxley*, 1760, pencil, chalk and watercolour.

shows two beech trees growing on a pronounced knoll or tump, with a curving path leading to a distant tower. To the left is a pollarded tree while to the right is a rustic fence. The 'remote paths and rickety fences dominated by the majestic old trees, seem to anticipate Uvedale Price's later theories of the Picturesque'. There is also a clear contrast of form between the recently cut pollarded tree to the left and the two trees on the tump.[14]

Pollards appear very frequently in many of Gainsborough's later paintings and this frequency can be put down to several factors. First, it demonstrates the important influence of the Dutch landscape paintings of Ruisdael and others, which we have seen were a crucial early influence on Gainsborough. Second, it probably reflects their very common occurrence along roadsides and hedgerows in the English landscape. Most trees adjoining fields would be pollarded, so their regular appearance in Gainsborough's work in that sense reflects the real landscape. Another key factor is that they often form vital framing devices for his landscapes. In *Sunset: Carthorses Drinking at a Stream* of *c.* 1760, for example, the shape of the dead pollard helps to reinforce the circular structure of the painting, allowing a view through to a distant landscape. In portraits such as *William Poyntz* (1762) and others, the gentlemen lolling against pollarded trees demonstrate their close connection with rural life. But the trees also form a useful dark contrast to the portraits and a framing device. The portrait of William Poyntz also emphasizes the regrowth of young leaves from the heavily cut young willow pollards in the middle distance.[15]

Two other very significant influences on Picturesque aesthetics were the Italian Salvator Rosa (1615–1673) and Claude Lorrain (Gellée, 1604–1682), who was born in Lorraine but lived most of his life in Rome. These two painters were among the most sought-after artists in British collections in the eighteenth century and were a strong influence on English artists such as Turner and Constable. Trees were crucial to both painters in providing contrasts of light and shade, in giving form and structure and in provoking different atmospheres, but neither painted

46 Thomas Gainsborough, *William Poyntz of Midgham and His Dog Amber*, 1762, oil on canvas.

47 Salvator Rosa, *Study of Trees*, 1640s, ink drawing.

trees whose species can readily be identified. In Rosa's painting *Mercury and the Dishonest Woodsman* (c. 1663; illus. 70), for example, which depicts Aesop's fable of the importance of honesty, the dark trees in the centre of the painting provide a mysterious and gloomy backdrop, while the jagged and broken trees frame the painting and give the landscape a 'brilliant structure'. The trees are not identifiable, and there is no evidence of the quotidian tasks that the woodsman of the title had been performing with his axe; rather the trees appear to have been damaged by storm or tempest. Uvedale Price bought six drawings by Salvator Rosa which he considered to be 'admirable specimens of the unparalleled freedom & lightness of his pen' in 1768, when he was in Perugia on his grand tour. These included a *Study of Trees, Study of Stump of Old Tree, A Magnificent Study of a Tree* and *Four Monks Seated at the Foot of a Tree*. The *Study of Trees* shows extraordinarily shattered and contorted stems characteristic of trees damaged by an ice storm or 'galaverna', which epitomize aspects of the Picturesque later to be theorized by Price.[16]

🌿 Picturesque aesthetics and trees

Picturesque aesthetics became one of the most potent and influential ways of interpreting and understanding landscapes and trees from the eighteenth century onwards. The historian David Watkin considers that 'the theory and practice of the Picturesque constitute the major English contribution to European aesthetics' and that between 1730 and 1830 the 'Picturesque became the universal mode of vision for the educated classes'.[17] And yet the origin of the term and its varied meanings are difficult to pin down. In the 1750s two works were published which, while not using the term 'Picturesque', stimulated considerable debate. William Hogarth's *Analysis of Beauty* (1753) argued that the eye preferred variety and intricacy to symmetry and that 'those lines which have the most variety themselves, contribute towards the production of beauty.' His 'line of beauty' was not simply a curved line but a 'precise serpentine line' which was neither 'too bulging nor too tapering'.[18] This was followed a few years later by Edmund Burke's enormously influential *Philosophical Enquiry into the Origin of Our Ideas of the Sublime and Beautiful* (1757).

When Horace Walpole published *The History of the Modern Taste in Gardening* in 1780 as the last volume of his *Anecdotes of Painting in England*, he made the link between painting and landscaping explicit. Tying the new style of gardening directly to his project of encouraging English painting, he asked that if 'we have the seeds of a Claud or a Gaspar amongst us, he must come forth. If wood, water, groves, vallies, glades, can inspire poet or painter, this is the country, this is the age to produce them.'[19] Walpole traced the line in the development of modern gardening from Charles Bridgeman, who 'first thought' of the 'capital stroke' of the 'sunk fence', through to William Kent, whose 'great principles' were 'perspective, and light and shade'. Kent's 'Groupes of trees broke too uniform or too extensive a lawn; evergreens and woods were opposed to the glare of the champain', and by 'selecting favourite objects, and veiling deformities by screens of plantation', he was able to realize 'the compositions of the greatest masters in painting'. Walpole noted that some of the newly created gardens and parks were directly comparable to the greatest art. At the Earl of Halifax's Stansted, for example, he relished that 'great avenue' which traversed 'an ancient wood for two miles' and the 'very extensive lawns at that seat, richly inclosed by venerable beech woods, and chequered by single beeches of vast size'.

He was particularly taken by the view from 'the portico of the temple', where 'the landscape that wastes itself in rivers of broken sea, recall such exact pictures of Claud Lorrain, that it is difficult to conceive that he did not paint them from this very spot.' Walpole approved of much of Lancelot 'Capability' Brown's landscaping, noting that Kent had been 'succeeded by a very able master', although he could not name him, since 'living artists' were not included in his history. Although he was concerned that 'the pursuit of variety' threatened the modern style of gardening, he stressed that 'In the mean time how rich, how gay, how picturesque the face of the country!' and that there had been so much improvement brought about by the modern style of gardening that 'every journey is made through a succession of pictures'.[20]

William Gilpin (1724–1804) was the most important and influential writer to spread the idea of the Picturesque in the second half of the eighteenth century, through the publication of his *Essay on Prints* (1768) and his subsequent tour journals, which offered observations on 'picturesque beauty'. Gilpin, who had a combined career as a schoolmaster, vicar, philanthropist, scholar and artist, had first developed his views on the Picturesque in an early essay, *A Dialogue upon the Gardens of the Right Honourable the Lord Viscount Cobham at Stowe* (1748), which characterized it as 'a term expressive of that peculiar kind of beauty, which is agreeable in a picture'.[21] Gilpin argued in his *Essay on Picturesque Beauty* (1792) that certain qualities present in nature – roughness and ruggedness, variety and irregularity, chiaroscuro – could combine to form 'picturesque beauty', a phrase that Gilpin admitted was 'but little understood', but by which he meant 'that kind of beauty which *would look well in a picture*'.[22] The fullest exposition of Gilpin's concept of the Picturesque came from his various tour journals: on the River Wye (1770; published 1782), the Lake District (1772; published 1786), North Wales (1773; published 1809) and Scotland (1776; published 1789). Gilpin's ambition on these tours was to 'examine the face of a country *by the rules of picturesque beauty*: opening the sources of those pleasures which are derived from the comparison'.[23] The manuscript versions of Gilpin's tour journals received wide circulation among an influential group of friends and acquaintances including Thomas Gray, William Mason, the Duchess of Portland, Mrs Delany and Queen Charlotte, to whom the volume on the Lakes was dedicated when it was published in 1786.

The most careful elaboration of Picturesque aesthetics at the end of the eighteenth century was Uvedale Price's *Essay on the Picturesque* of 1794, which aimed 'to shew . . . that the picturesque has a character not less separate and distinct than either the sublime or the beautiful, nor less independent of the art of painting'.[24] For Price, the study of nature and the works of great artists – such as the landscapes of Claude, Salvator Rosa and Nicolas Poussin as well as those of Dutch and Flemish artists such as Ruisdael – was essential in understanding how to design and lay out grounds and estates. His emphasis on connection and local knowledge implicitly validated the authority of informed landowners (such as Price himself) as those best placed to effect changes to the landscape. 'He therefore, in my mind,' he wrote, 'will shew most art in improving, who *leaves* (a very material point) or who creates the greatest variety of landscapes'. He introduced a third category of 'picturesqueness' which stood for ruggedness, variety and character. According to Price, 'the picturesque fills up a vacancy between the sublime and the beautiful, and accounts for the pleasure we receive from many objects, on principles distinct from them both.' This allowed him to defend aspects of the landscape that he saw were being cleared away by the mania for improvement: 'old neglected bye roads and hollow ways', 'old, mossy, rough-hewn park pales', rustic hovels, mills and cottages. He was particularly fond of old pollards that resulted from the 'indiscriminate hacking of the peasant' and the 'careless method of cutting, just as the farmer happened to want a few stakes or poles'. He celebrated the 'spirit of animation' found in 'the manner in which old neglected pollards stretch out their limbs across these hollow roads, in every wild and irregular direction' and found them in many ways more attractive than 'the finest timber tree, however beautiful' in terms of health and vigour.[25]

Forest scenery

William Gilpin was a successful schoolmaster at Cheam for several years. One of his pupils was William Mitford, who owned an estate at Exbury in Hampshire, was a verderer of the New Forest from 1778 and later wrote a major *History of Greece*. It was he who encouraged Gilpin to become vicar of Boldre in the New Forest in 1777, a post that produced an income of £600 a year. Gilpin dedicated his lengthy *Remarks on Forest Scenery* to Mitford in 1791. Gilpin was fascinated by the ancient trees of the New Forest, and on his first visit there exclaimed in a letter to the poet William Mason,

Such Dryads! Extending their taper arms to each other, sometimes in elegant mazes along the plain; sometimes in single figures; & sometimes combined! What would I have given to be able to trace all their beauteous forms on paper!! Alass! My art failed me. I could only sketch: and a sketch amounts to no more, than, N.B. Here stands a tree.[26]

Robert Mayhew has argued that many critics have underestimated the moral seriousness of Gilpin's Picturesque and emphasizes that his theological position as a Low Church Latitudinarian, in which he used 'the natural world as a mode of evidence that fitted their need for un-controversial proofs of God and Christianity', was of crucial importance in informing his ideas of the Picturesque.[27] Latitudinarians such as William Paley, in his enormously popular *Natural Theology*, first pub-lished in 1803, reinvigorated old arguments that the form and structure of nature were proofs of the existence of a creator. Trees and woods were often used to elucidate the argument. The botanist and theologian John Ray's description of English plants *Catalogus plantarum Angliae*, published in 1670, emphasized that there was a divine purpose in the creation of different plant species. He later stressed in *The Wisdom of God Manifested in the Works of the Creation* how the earth was 'curiously clothed and adorned with the grateful verdure of Herbs and Stately Trees, either dispersed or scattered singly, or as it were assembled in Woods and Groves'.[28]

Gilpin rarely drew attention to his theological beliefs in his tours, but in a note to his *Remarks on Forest Scenery* of 1791 he creates an allegory of an acacia tree which grew in his garden at Boldre in Hampshire. 'As I sat carelessly at my window, and threw my eyes upon a large acacia, which grew before me, I conceived it might aptly represent a country divided into provinces, towns and families.' The provinces were the larger branches, the smaller branches the towns, and the 'combinations of collateral leaves' represented the families made up of individuals. It was autumn and 'As I sat looking at it, many of the yellow leaves' were 'dropping into the lap of their great mother'; this was an 'emblem of natural decay'. A breeze caused many leaves to fall which 'might have enjoyed life longer. Here malady was added to decay.' The 'blast increased' and a 'shower of leaves covered the ground' as when 'pestilence shakes the land'. After the storm a few remaining solitary leaves survived as an 'emblem of depopulation'. Gilpin concludes by emphasizing that

'Nature is the great book of God' and that while the 'heathen moralist' knew that 'men, like trees, are subject to death', the 'same God presides over the natural, and moral world', and 'that power which revives the tree, will revive thee also'.[29]

He claimed that it 'is no exaggerated praise to call a tree the *grandest*, and most *beautiful* of all the productions of the earth'. While there was much beauty amongst the smaller flowers and shrubs, these could not compete in terms of '*picturesque beauty*' with trees whose 'form and foliage, and ramification' brought about 'the arrangement of *composition in landscape*' and received the '*effects of light and shade*'. Moreover, although he did not wish to 'set the tree in competition' with animal life to which 'we give the preference on the whole', he noted that 'every animal is distinguished from its fellow, by some little variation of colour, character or shape', while the larger parts, such as the body and limbs, are generally similar. With trees, however, 'it is just the reverse', and the smaller parts such as the leaves, blossom and seed are the same in all trees of a species, 'while the larger parts, from which the *most beautiful varieties* result, are wholly different'. But he saw as much difference in the beauty of trees as in human figures: while some were elegant because of their 'harmony of parts' and 'ease and freedom', in others 'The limbs . . . are set awkwardly; their trunks are disproportioned; and their whole form is unpleasing.'[30]

Gilpin was particularly concerned that landscapes which were chiefly dependent on woodland scenery were always 'open to injury' compared to those which depended 'on rocks, mountains, lakes and rivers'. The problem was that any 'graceless' hand could fell a tree and the 'value of timber' was its misfortune. He pointed out that 'in a cultivated country, woods are considered only as large corn-fields; cut, as soon as ripe'; when the timber was fit for particular purposes and uses, ''tho we may lament, we should not repine.' In the New Forest, for example, 'the vast quantities of timber, which are felled, every year, for the navy' meant that Gilpin's description was 'not the description of what it was in the last century, nor of what it will be in the next'. On the other hand, it was morally reprehensible when trees were cut prematurely 'to make up a matrimonial purse, or to carry the profits of them to the race-grounds, and gaming houses'. In these instances Gilpin wished that 'the profligate possessors had been placed, like lunatics, and idiots, under the care of guardians, who might have prevented such ruinous, and unwarrantable waste'. He argued that 'it is a much easier business to *deform*, than to *restore*', but there was always hope that 'as young trees are growing old, nature is also working up new *fore-grounds* to her landscapes.'[31]

There was an inherent conflict between some aspects of Picturesque aesthetics and the regular felling and cropping of trees. Gilpin lamented the 'capricious nature of *picturesque ideas*', many of which celebrated attributes 'derived from the injuries the tree receives, or the diseases, to which it is subject'. He demonstrates the changes in appreciation of trees by quoting 'a naturalist of the last age', William Lawson (1553/4–1635), the vicar of the village of Ormesby in Yorkshire, who wrote 'the first published work on gardening in the north of England, and . . . the first horticultural work written specifically for women'. Gilpin quotes Lawson's enumeration of the defects of trees:

> How many forests, and woods, says he, have we, wherein you shall have, for one lively, thriving tree, four, nay sometimes twenty-four, evil thriving, rotten, and dying trees: what rotten-ness! what hollowness! what dead arms! withered tops! curtailed trunks! what loads of mosses! drooping boughs, and dying branches, shall you see every where.

But Gilpin goes to argue that 'all these maladies, which our distressed naturalist bemoans with so much feeling' were now often recognized as 'capital sources of picturesque beauty' in both 'wild scenes of nature' and 'artificial landscapes'.[32]

He felt that for examples of the use and beauty of the withered top and curtailed trunk, 'we need only appeal to the works of Salvator Rosa', who often used 'the trunk of a tree in his fore-grounds', where a complete tree in its 'full state of grandeur, would have been an incumbrance'. Gilpin argues that 'ruins' of 'noble' trees are 'splendid remnants of decaying grandeur' which 'speak to the imagination in a stile of eloquence which the stripling cannot reach'. Their great age allows them to 'record the history of some storm. Some blast of lightening, or other great event, which transfers it's [sic] grand ideas to the landscape', and moreover in the 'representation of elevated subjects assists the sublime'. Gilpin felt that the blasted tree was 'almost essential' as a source of beauty for some natural and artificial scenes. Thus when a 'dreary heath is spread before the eye, and ideas of wildness and desolation are required', he asked 'what more suitable accompaniment can be imagined, than the blasted oak, ragged, scathed, and leafless; shooting

48 William Gilpin, *An Unbalanced Tree Bending over a Road*, ink and wash drawing. The drawing was etched by Samuel Alken for Gilpin's *Remarks on Forest Scenery . . .* (1791).

it's [sic] peeled, white branches athwart the gathering blackness of some rising storm'.[33]

Gilpin's Picturesque way of seeing also focussed on 'nature's minutiæ', such as moss growing on trees, which 'touches not the great parts, *composition* and *effect*' but is a 'beautiful object of imitation' in 'coloured landscape'. On his walks around the New Forest he had 'often stood with admiration before an old forest-oak, examining the various tints, which have enriched it's [sic] furrowed stem'. Near the roots he often found the 'green, velvet moss, which in a still greater degree commonly occupies the bole of the beech'. Higher up the trunk 'you see the brimstone colour taking possession in patches', one kind smooth and the other of which 'hangs in little rich knots, and fringes'. In addition 'you often find a species perfectly white' and there are also 'touches of red; and sometimes, but rarely, a bright yellow, which is like a gleam of sun-shine'. Gilpin referred to all these 'excrescences' as mosses but admitted that 'those particularly, which cling close to the bark of the trees, and have a leprous, scabby appearance, are classed, I believe, by botanists, under the name of *lychens*: others are called *liver-worts*.' But whatever they were called, Gilpin felt they added 'great richness to trees' and that the painter admired them 'among the picturesque beauties of nature' while 'the wood-man . . . brushes them away.' Uvedale Price agreed and added them to his catalogue of the features of ancient trees: 'the deep hollow of the inside, the mosses on the bark, the rich yellow of the touch-wood, with the blackness of the more decayed substance', which caused a Picturesque 'variety of tints, of brilliant and mellow lights, with deep and peculiar shades'.[34]

A key theme of Picturesque woodland management was to open up views through woodland and framed by trees, as in a painting by Claude. Here the trees were not so much the focus of attention as a means of dis-playing landscapes. Uvedale Price, for example, managed his old coppice woodlands and plantations carefully to open up the best views of the surrounding countryside. He enhanced the views available from the rides around the estate and framed what he thought of as a gallery of pictures. He also worked *within* his woods and coppices to make com-positions and pictures. In 1796 he told his friend Lord Abercorn how he was 'clearing some parts among the shrubs, & making glades & openings on a small scale'.[35] Over twenty years later, in 1818, he was still improving these landscape views and using the analogy that in every block of marble there is a fine statue; it only needs to have the rubbish removed from around it. He told Lord Abercorn that at Foxley the blocks of marble were three unprofitable coppice woods which contained 'treasures of beauty' including 'fine timber trees' and 'a number of old yews, thorns, nuts, hollies, maples'.[36]

Price 'disguised the lines' of his new paths, walks and rides by thin-ning and pruning his trees with 'a proper mixture of caution & boldness', stressing that this was 'at least as necessary as planting'. To help in this job he trained one of his workers to be 'a pruner who gets up into the very highest trees (not from my teaching however) & perfectly compre-hends + executes my ideas'. The result of this vigilance was that 'single trees & groups' which before were '*uniformly* heavy & massy' were made 'much more varied, light & airy'. In this way he tried to 'apply to nature' the principles of the art of which he was such a connoisseur. This prun-ing and thinning he found to be 'a source of great interest & amusement both at the time & afterwards'.[37]

Unnatural practices

For Gilpin, all trees which had forms 'that are *unnatural*, displease'. He found trees 'lopped into a may-pole, as you generally see in the hedge-rows of Surry' to be 'disgusting'. Clipped trees such as yews, lime hedges and pollards were disagreeable because the trees were unnatural and disproportioned and their branches joined the trunk awkwardly. An overgrown pollard could sometimes, however, 'produce a good effect, when nature has been suffered, for some years, to bring it again into form'. But when trees were repeatedly cut for leaf fodder they could only be ugly. Ash trees in the New Forest, for example, had 'leaf and rind' which was 'nutritive to deer' and this 'disagreeable circumstance' meant they were heavily browsed in the summer and became 'mangled, and deformed'. Very occasionally an individual pollard could achieve beauty through peculiar and ephemeral circumstances. Gilpin remembered that once 'in autumn I have seen a beautiful contrast between a bush of ivy, which had completely invested the head of a pollard-oak, and the dark brown tint of the withered leaves, which still held possession of the branches.'[38]

Another common example of a displeasing 'unnatural' form was where 'some single stem was left to grow into a tree' from a pollard. Here Gilpin complained that 'the stem is of a different growth; it is dispro-portioned; and always unites awkwardly with the trunk' (illus. 49). His voicing of concern over the ghastly effects of unnatural pollarding and shredding (in which side branches were removed) was added to an already loud chorus of attack. One of the strongest opponents was Arthur Young (1741–1820), Secretary to the Board of Agriculture, who found pollarding and shredding barbarous and vile practices which ruined timber and destroyed the landscape of southeast England: 'The beauty of all this country is wretchedly hurt by the abominable custom of stripping up all trees; in so much that they look like hop-poles.' The strong regional variation in the frequency of pollarding was noted: he did not find this 'detestable practice' in Norfolk, Suffolk and parts of Essex. The 'barbarous practice' of converting 'timber trees into pollards' was strongly condemned. Many Middlesex hedges were 'disfigured by pollard trees' which were 'rotten' and the shredded trees were 'stripped of their side-branches, like May-poles'. The rise of the Picturesque movement meant that regularly cropped trees were seen as anachron-istic adjuncts to agriculture while old, overgrown pollards became

49 William Gilpin, *A Pollard on which a Single Stem is left to Grow into a Tree*, wash drawing. This drawing too was etched by Samuel Alken for Gilpin's *Remarks on Forest Scenery . . .* (1791).

celebrated. Uvedale Price told the landscape improver Humphry Repton in 1792 that 'he would deserve a statue if he could inspire Mr Pitt with such an aversion' to stripped elms 'as to make him exert his great authority & eloquence to put an end to such a horrid practice'. He argued that the ending of such a practice 'would do more towards beautifying the face of England than all the sums that ever have, or will be laid out in improvements'.[39]

Unnatural pollards and shredded trees continued to be castigated for their ugliness and poor form in the nineteenth century and there was debate as to whether a tree could ever recover its beauty once pollarded. In 1811 the Suffolk landowner Revd Sir Thomas Gery Cullum considered that 'trees in many parts of England' were 'like a cabbage stuck on a May pole, or left with long stumps like the teeth of a rake'. Sir Henry Steuart found it unlikely in 1828 that a tree that had been cut in this way could ever recover 'its natural and free conformation'. The artist Jacob Strutt in *Sylva Britannica* (1830) found willows 'disagreeable to the eye of the painter', especially after pollarding, when 'their decapitated trunks then present an unsightly spectacle, not much improved when they again sprout forth.' James Main argued in 1839 that dismemberment was akin to 'mutilation', 'distortion' and 'bad taste', and that 'only natural disbranchement by wind or lapse of time' could produce a natural shape that could be admired.[40]

The tension between a liking for ancient pollards and a hatred of recently cropped ones is quite striking in the middle years of the nineteenth century. As the practice became less common, abandoned and overgrown pollards became increasingly valued for their Picturesque qualities. William Craig found in 1821, for example, that 'Lopping and pollarding also produce wonderful changes on the aspect of trees, sometimes rendering them highly picturesque, and sometimes disgusting; but always disproportioned from their natural character.' The author of *Woodland Gleanings* (1865) found that 'pollards, being rendered

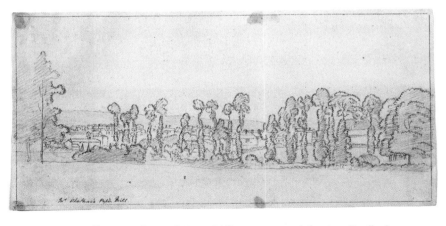

50 Paul Sandby, *Mr Whatman's Paper Mill*, c. 1794, pencil drawing. Sandby has carefully drawn a row of trees, probably elms, that have been shredded with fresh foliage growing from their trunks.

unnatural in form, are disagreeable; though sometimes a pollard produces a good effect, when nature has been suffered, after some years, to bring it again into shape.' By the mid-nineteenth century authors were frequently alluding to the historical associations of ancient pollards at sites such as Burnham Beeches and Sherwood Forest. The author of *Gleanings in Natural History*, writing of the pollards of Windsor Forest, was 'imperceptibly carried back to the many interesting historical facts which have happened' in their lifetime: 'I can fancy that our Edwards and Henrys might have ridden under their branches.' The aesthetic view that regularly cut trees were unsightly and exploitative had become firmly entrenched in late nineteenth-century Britain. This Picturesque sensibility eventually made pollarding a potent symbol of forest over-exploitation and the practice was seen as incompatible with early ideas of conservation.[41]

FIVE
Pollards

MANY trees in gardens, streets and orchards are regularly pruned, pollarded and shaped, while those in the countryside are generally left to grow untouched. It was very different in the past, when most trees growing in hedgerows and on common land throughout Europe and in many other parts of the world were pollarded or shredded for the production of leaf fodder and firewood. There is no doubt that pollarding and shredding are very ancient practices. In prehistoric Europe elm and ash leaves from pollarded trees were of enormous importance as feed for animals. This was particularly true in areas or periods in which the availability of grass, in the form of pasture or hay, was limited: for example during the winter or in periods of drought. Most tree species were lopped for leaf fodder and fuel. The elm, lime, ash, oak, alder, rowan, hazel, hawthorn and some conifers, such as silver firs in the Apennines and northern Greece, were cut as fodder. This is a largely forgotten history: forest historians have until recently tended to ignore the practice because it was not seen as part of forest history; conversely agricultural historians did not consider it part of agriculture. Pollarding was thus of no more than peripheral interest to both groups. This lack of interest has masked the importance of tree management in agricultural and pastoral areas.

The value of the products of pollards is shown by the complex mix of laws and rights over different parts of trees and by disputes over who had rights to cut branches or take crops from trees. In present-day Crete, for example, the rights to the acorns of certain oak trees can still be divided between several different farmers. In England a legal distinction can be made between timber trees, usually owned and managed by the landowner, and other trees, which could be pollarded by their tenants. This encouraged landowners to reduce the number of trees that could be pollarded and there was a sustained campaign over 200 years or so by many agriculturalists and foresters to halt the practice. While there has been a revival of interest in pollarding and old trees, which are increasingly valued as important features of cultural landscapes and are included in conservation programmes, in most Western countries pollarding is nowadays almost a forgotten practice. This chapter uses examples from Greece and Britain to examine the collapse of this traditional form of management.[1]

Leaf fodder

Many animals have evolved to eat the leaves of trees and shrubs and complicated ecological relationships have developed over centuries as a consequence. Before humans started to domesticate animals, the vegetation of many parts of the world was shaped and modified by the browsing and grazing of many different species.[2] With the domestication of animals such as sheep, goats and cattle, the large, mobile flocks and herds produced began to exert an even greater effect on the vegetation. With contemporary modern farming techniques the interrelations between animals and vegetation are often disconnected, but for thousands of years the landscapes of many areas were created by the complex interplay of pastoralists, their animals and the vegetation they produced. Trees and shrubs, through the provision of leaf fodder, played a crucial part in the animal and human economy and the various techniques employed in cropping and preserving leaves have shaped both trees and our perceptions of them.

One of the most common, and long-standing, methods of management was to cut the branches of trees off at about head height. This had three main benefits. First, as long as the branches were cut above the reach of the thrusting tongues of cattle and goats, the tree would send out new shoots that could be cropped at some future time. Second, regular and fairly frequent pollarding meant that the branches to be cut were relatively thin and hence easy to cut with a sharp, iron hook. Third, by cutting at head height, the cropping could be undertaken without having to climb high up into the branches of tall trees. Such pollarding was thus a simple, renewable way of harvesting leaf fodder. There was no fixed method of cutting branches; sometimes branches were lopped all the way up a stem in a system often called shredding, but sometimes odd branches were cut whenever needed. Trees that were regularly cut carried on producing a regular and healthy crop of leaves every few years. Moreover, the branches that were also produced could be put to a wide range of uses, especially the making of faggots for fires. If the branches were allowed to grow for a few more years, depending on the species, they could also be used for construction, making tools, fencing and firewood.

Pollarding and shredding of trees have been widespread and common practices in many parts of the world, including Europe, Japan and the tropics. In Britain they remained important until the eighteenth century but by the mid-nineteenth century were becoming increasingly rare and had virtually disappeared by the mid-twentieth century. In contemporary Europe pollarding is undertaken to provide fodder in the mountains of Greece and also takes place in mountainous parts of Spain and in Bulgaria. In parts of northern Italy ash, alder, poplar and beech are still used for fodder, directly consumed or stored in barns (illus. 71). The poles are used for firewood and fencing. I still remember the sight, in 1985, of a cyclist confidently yet precariously carrying several poles he had just cut as he rode up one of the steeper roads in the Euganean hills near Vicenza. Rows of freshly pollarded trees can still be glimpsed from the Autostrada as they stretch out across the Po Valley. In Norway the branches of poplars, ash, elms, pine, goat willow and rowan were pollarded, dried and stored in the past and in western Norway a few farmers continue to use leaf fodder because they consider it good for the health of their animals. The Finnish botanist Carl-Adam Hæggström describes pollard meadows which produce hay and loppings along the shores of the Baltic in southwest Finland and in the Åland Islands. Pollards are also still used for producing firewood, charcoal and constructional wood. In France the demand for firewood encourages pollarding, but, in contrast with the ancient practice, the branches are cut at longer intervals to produce suitable logs.[3]

Recent studies have emphasized the enormous importance of the practice for rural populations in tropical areas. The important research of the French geographer Sandrine Petit and others has shown that tree fodder is crucial for livestock bred extensively in dry areas like the Sahel. The nutritional value of tree forage is recognized by the Food and Agriculture Organization and this resource is so important that it is traded and pollarding is regulated locally. For example, in Mali the fodder of *Pterocarpus erinaceus* has a ready market in Bamako, the capital city. In 1989–90, 1,400 tons of fresh fodder were sold in Bamako.[4] The lopping and pollarding of trees is a common practice in Asian mountains and fodder trees are cultivated by farmers in India, Nepal, Sri Lanka and other countries. In Rajasthan trees are appropriated for fodder collection and the rights to exploit a tree are the object of negotiation and trade. In Nepal there is deep local knowledge about the different nutritional values of the various fodder trees. Near Kathmandu the increase in population has encouraged the planting of trees close to houses to be lopped for fodder, litter and fertilizer. Pollarding is common in agriforestry systems, such as the traditional Quezungual system in Honduras, where the naturally regenerated trees are retained and regularly pollarded when land is cleared for cultivation. These examples are not exhaustive but reveal the importance of pollarding in the systems of production of rural areas. In addition to fodder, the leaves of trees serve as fertilizers and trees are pollarded so the branches can be used for fuel wood, to make tools and in building construction.[5]

Leaf fodder in Greece

In the mountainous parts of Greece grazing is still an important part of the local economy and the vegetation is heavily browsed by large flocks and herds of sheep and goats. Every tree and shrub is nibbled and the closely cropped vegetation is a continuum of shrubs and trees, many of which, although only a few feet high, are many years old. At Kasteneri in northern Greece in 2007, for example, I saw hornbeam and juniper trees so closely grazed that they formed a green topiary carpet of heavily browsed trees, which only occasionally grow above the height that can be reached by the animals. The leaves that grow above the browsing line are too valuable as fodder to be left alone, and for many hundreds of years these leaves have been cropped. One way of gaining an understanding of the practicalities and art of pollarding is to interview those people who still pollard, or did so when they were younger. The Greek forest historian Eirini Saratsi undertook an important study of this kind in the Zagori region of northwestern Greece, in the Pindus mountains not far from the Albanian border. She interviewed many of the older residents to discover how they used trees in the past. Nowadays, following a lengthy period of rural depopulation since the Second World War, much of the woodland in the region has naturally regenerated over areas which were formerly cultivated and grazed. This woodland is valued as a habitat for bears and wolves, and traditional forms of management are discouraged so that these animals might thrive. Before that war, however, the woods were used in three main ways: for firewood, fodder for animals and timber. Goats were allowed to graze in some areas, whereas wood pastures were usually devoted to sheep grazing. Leaf fodder was provided by individual trees that were left to grow along the edges of fields. The evidence for the importance of leaf collection comes from legal documents, the shape and form of surviving trees and the testimony of people who engaged in the practices.[6]

As in many mountainous parts of the Mediterranean, the upper slopes in Zagori were traditionally grazed by nomadic or transhumant flocks and herds while the lower slopes near villages were often terraced and cultivated. When walking around villages such as Micro Papigo, which nowadays survive largely through green tourism, it is possible to spot many old pollarded or shredded trees growing along the edges of fields, now largely abandoned, or in narrow strips of woodland, known locally as *kladera*, which were used especially for the provision of winter fodder. The pattern of small woods and fields was determined to some extent by the topography. The fields were made and cultivated on the less steep parts of the slope and the steepest, often very rocky parts were left as woodland. The woods were a vital source of fodder for the villagers, who, unlike the nomads, were not able to travel away from the mountains in search of winter pastures.

By far the most common type of tree growing in the *kladera* was oak; Eirini Saratsi found eight species of oak growing in Zagori, and all were used in one way or another for feeding animals. One particularly useful characteristic of the deciduous oaks is that when cut in the summer, the branches retain their leaves when dried, which makes them easier to store. The different species of oak had characteristics which were well known to the people who fed the animals. A 70-year-old man said 'Here we have Tzero [*Quercus cerris*], we have Drios [*Q. frainetto*], we have Douskou [*Q. robur*] we have Granitsa [*Q. petraea*] . . . Tzero, we did not cut it very much because it crushes into bits easily, and was also a little sharp.' A 90-year-old woman recalled that the evergreen prickly oak was used differently: 'we did not store it indoors, we used to go even when it snowed . . . to bring a branch for the goats to eat.' The branches of other species of tree which goats preferred, such as the field maple, hop hornbeam and hornbeam, did not retain their leaves once cut, so were used only as fresh leaf fodder in the spring. The species most favoured for young goats were the flowering branches of the lime tree (*Tilia alba*), locally called *lipanthia*; a man from Micro Papigo noted that 'we used to cut it a lot but the problem was where could we find it?' The selection of the correct type of tree for pollarding and shredding required detailed and thorough knowledge.

There were also different strategies for the frequency of cutting individual trees. In some cases a tree might be cut every year, while others might be left to grow for three or four years. One 69-year-old man celebrated the appearance of the pollarded trees:

> You cut it this year, and the next year it grew big, so you cut it again . . . Oh you should see how beautiful they were. Because every year they were cut they become thick and the same again and again . . . people used to cut the top as well, of course they cut it because it was growing very tall . . . They became round, can you imagine how beautiful they were? I have cut, ooooh, maaaaaaany loads.

In other places people remembered that they cut the trees every other year, or half the branches off a tree one year, and the other half the next. One man recalled 'you climbed the tree and pruned all the branches, you only left the stem and one or two shoots at the top. The rest of the tree was pruned . . . Every three, every four years and it was growing again.' Clearly the exact practice varied from village to village depending on the type of trees and the uses to which the branches were put.[7]

Recent research by the Cretan forester Pantelis Arvanitis in a detailed study of the history of woodland management practices in the Psiloritis mountains of central Crete has shown the importance of different tree species for the provision of leaf fodder for shepherds. The Cretan word used for the whole process of collecting leaves and feeding the sheep and goats in Psiloritis is *kladizo*, meaning 'branching'. The most important leaf fodder trees were the maple (*Acer criticum*) and the kermes oak (*Quercus coccifera*); these are also the most common trees growing in the Psiloritis mountains. Other trees, such as *Q. ilex* and *Phillyrea latifolia*, were not used because shepherds considered that their animals did not like to eat them. The cutting of branches was seen as beneficial to the trees. One shepherd remembered that 'Many times when someone had a few goats, 100 animals, he could cut now and then kermes oak in the end of the summer, autumn, and that was good for the trees.' A local mayor emphasized that the cutting did not damage the trees. The shepherd did

51 Grazed woodland, Zaros, Crete, 2010.

not 'go to cut kermes oak, to destroy it. He will cut specific branches, he was pruning it and then it would sprout again later. It was revivifying the tree and that was very important.' He remembered that they would say 'I am going to *kladiso* the sheep. When they say *kladizo*, they would climb in a tree with a saw that the shepherd had with him, and he would cut three, four, branches, no more from each tree.' The branches fell to the ground and the animals rushed over to eat the fresh foliage. Some shepherds reported that the animals so much preferred this lush food that they would run as fast as possible to the tree where they were cutting branches as soon as they heard the action of the saw or axe.[8]

It is possible that the dominance of the many shrubby maples and kermes oaks in the area is due to careful protection by shepherds over many centuries as it was in their interests to ensure that there was a regular supply of fodder. On the other hand, the survival of some areas of large old holm oak (*Quercus ilex*) near the village of Zaros was ascribed by one shepherd to the unpleasant flavour of its leaves: 'goats do not eat it as they eat the shrubby kermes oak. Kermes oak although it has spines goats like to eat it. It is sweet, although it has thorns.' *Phillyrea latifolia* and holm oak 'have no thorns but they do not eat them, they are avoided. I am saying that both are bitter. That's the reason that holm-oak survived, it is in the mountain, the side of Zaros village, very big trees.' Others argued that the young leaves of *Q. ilex* were palatable, and that older trees survived on cliffs or other areas which sheep and goats were not able to reach.

The shepherds agreed that the most important fodder tree was the maple: it was favoured by their animals and the leaves have no spines. Branches were cut throughout the summer period when the flocks and herds were in the mountains. The best time for cutting, however, was in July, when the maple seeds had formed and the foliage was at its optimum extent. One 78-year-old shepherd pointed out that the best time to cut the foliage was linked with the life cycle of their stock. 'When the tree had both leaves and fruits, we cut it in order to fatten the young goats . . . And the ones that were planned to be killed, we feed them as well, when the fruits of the tree were matured.' The kermes oak was shredded from August onwards, when there was much less grass, and especially in September and early October, when the oak branches bore acorns as well as leaves. Later the acorns would be threshed from the trees to provide feed for the goats, and in October and November the acorns would fall down from the branches themselves, especially in heavy rain. Cutting the oak branches was not necessarily an easy job. Shepherds had to be agile to be able to climb up into the trees, and the spiny oak leaves meant that clothing was often damaged: one shepherd wryly commented that every time you cut leaf fodder you needed a new shirt. The leaf fodder could also be used as a management tool. Pantelis Arvanitis found that shepherds gave their stock leaf fodder when new animals were added to a herd, so that they would keep close to the shepherd's hut until they became used to their new home.[9]

But the branches do not always have to be taken to the animals; sometimes the goats climb into the trees themselves to browse the delectable fresh leaves. We saw clear evidence of this on a visit to the Forest of Zagori in July 2010. Many trees had fresh marks and abrasions on their trunks caused by hoofs where goats had climbed regularly into the low canopy, and where goats had browsed along the branches, they left characteristic flat areas of closely nibbled foliage. Several shepherds described this, noting that the goats climbed into the oak trees, but not every tree; it had to have the correct form of trunk. Once a goat had established a climbing route, others would follow, and all the leaves the goats could reach, even at full stretch, they would eat. In eastern Crete it has been reported that shepherds chopped small steps into the tree and placed stones in to ease the path for the goats. But this was not currently the case in Psiloritis. One shepherd stated that they used to 'put some stones so that we could reach the tree and thresh it' for acorns and that they then removed the stones. He was adamant that he did not want the goats to climb the trees and graze them because there was a danger that the goat would be killed. It was relatively common for a goat to slip in the act of reaching for the leaves, for its horns to get caught in the branches and for the goat to hang until it died. This was most likely to occur when the goats were grazing in trees well away from shepherds, and in the past, when there were more shepherds and the herds were more closely watched, such tree grazing was much more common. Certainly the ability of goats to climb trees and eat the leaves themselves would have reduced the need to cut branches; it is likely, however, that the two practices ran side by side depending on the availability of suitable trees and the number of stock and shepherds.

Leaf fodder in Crete, as in many parts of the Mediterranean, was an important supplement to grasses and other herb species, and still

52 Grazed oak tree, Zaros, Crete (left).

53 The trunk of an oak regularly climbed by goats, Zaros, Crete, 2010 (right).

provides essential nourishment, especially towards the end of the long, hot Cretan summers. These two recent Greek studies, one from the far northwest of the country and one from the deep south, emphasize the widespread nature of the use of leaf fodder for sheep and goats throughout modern Greece. Although the practice is now dying out, the testimony of retired shepherds and farmers provides fascinating insights into the specialized knowledge and techniques used to make use of leaf fodder. Moreover, modern foresters are beginning to realize that these old and indeed ancient management practices may be of vital importance in the development of management strategies for the conservation of old trees and associated habitats. This is true not only for Greece but for many of the countries encircling the Mediterranean.

Pollarding decline in Britain

In many parts of Britain old pollarded trees survive in fields, common land, parks and hedgerows as witnesses of the former importance of the ancient practice. They can be found in old hedges that predate the enclosure movement and are typical of ancient countryside character-ized by a high density of hedges, non-woodland trees and many small woods. Many tree species growing in Britain, including ash, elm, beech, oak, willow, alder, hornbeam, hawthorn, maple, blackthorn, holly and sweet chestnut, were pollarded (illus. 72). The main species that is still regularly cut in this way is the willow, and willow pollards are still a frequent sight along rivers and canals. The importance and frequency of the practice in the past can be explored by studying surviving ancient pollarded trees, landscape paintings, maps and written documents. The field evidence provided by the shape and form of old trees is an impor-tant clue. Ruth Tittensor in her fascinating study of the Mens, a wooded common in Sussex, notes that although pollarded beeches are very common, there is little mention of the practice in documentary sources. The frequency of pollarding depended on the product required: for fodder, trees were often lopped annually; for firewood or poles, the lapse was variable, depending on the species. Oliver Rackham has used tree-ring evidence to show a pollarding cycle of 13 years in Epping Forest, between 12 and 36 years in Hatfield forest and between 18 and 25 years at Hainault.[10]

In order to gain a broad picture of changing attitudes to pollards in Britain, Sandrine Petit and I examined around 200 published works by foresters, agriculturalists and gardeners from *c.* 1600 onwards. In general, at the beginning of the period, pollarding, although a relatively minor element in most texts, was at least mentioned and discussed. By the late nineteenth century, however, it was largely ignored. The widespread importance of the technique is indicated by the large variety of synonyms and related words, such as 'dodderel', 'dodder-tree', 'doddle', 'polly', 'pollenger', 'pollinger', 'stockel' and 'stoggle'. 'Shredding' and 'shrouding' are related terms which denote the removal of the side-branches, but both techniques could be performed on the same tree at different times. Terms used to designate trees from which branches had been cut included 'may-pole', 'hop-pole' and 'dottard'. Some terms had negative or anthropomorphic connotations, such as 'to mutilate' and 'decapitate'. Local names for pollards include 'greenhews' in the north and 'deer fall' in Wensleydale. The word 'rundle', used in Radnor, is equivalent to 'runnel' and both designate a pollarded tree, but especially a hollow one.[11]

In the seventeenth century the leaves of trees remained important in Britain for winter forage. Elm was particularly valuable: John Worlidge wrote in 1669 that branches and leaves of the elm were good for cattle and that in the winter cattle would eat elm before oats. He noticed that it was easy to shape the tree in any way one pleased and stressed that pollards and shrouded trees were well adapted for grazed areas because they could not be browsed by cattle. Pollarding was perceived differently depending on the tree species concerned. Many authors condemned the pollarding of oak, which was generally given a special status. The London nurseryman, gardener to the Earl of Essex and influential author Moses Cook argued in 1676 that 'our Yeomen and Farmers are too much subject to spoil such trees as would make our best Oaks, by heading them, and making them Pollards.' He thought there should be a strict law 'to punish those that do presume to head an Oak, the King of Woods, tho it be on their own land'. He argued also that oaks located in hedges should be shredded like elms so that they could grow up as timber. In contrast the influential author Batty Langley (1696–1751) recognized that the pollarding of oak was profitable: 'Oaks being headed, and made pollards, are in some countries very profitable, and will last for many Ages.' The pollarding of 'Aquatiks' such as willows and poplars was generally encouraged, although John Worlidge, writing in 1669, thought that if you let Aquaticks grow without 'topping' them 'they are then more Ornamental' but not economically 'beneficial'.[12]

Pollarding was extremely important for the production of firewood which was the main fuel for cooking and heating. The agricultural writer Arthur Standish argued that lopping trees was profitable: a landowner told him that 'he did every year loppe five to fiftie of tenne yeeres growing, the which wood he could yearly sell for fortie shilling' on land which if let normally would only produce an income of about 26 shillings. Worlidge encouraged tree planting in hedgerows: 'For Ash, Elm, Poplar, Willow, and such Trees that are quick of growth, it is a very great profit that is made of them where Fewel is scarce, by planting them in Hedge-rows, and other spare places, and shrouding them at five, six, eight or ten years growth.' The big advantage was that 'They are out of the danger of the bite of Cattle, and require no fence.' Special techniques were employed to help maintain a healthy and productive tree. John Mortimer, a merchant with an estate in Essex, advised in 1707 that elm trees should be shredded from the bottom upwards, taking care to preserve the top of the tree and leave it uncut. In this way a constant supply of firewood could be produced by the side boughs and the bodies of the trees could 'afterwards be good timber'. If trees were kept as pollards he recommended that they were cut every ten years.[13]

Most seventeenth-century authors were not opposed to pollarding, but they were concerned about its effect on the supply of naval timber. In the first edition of his enormously influential *Sylva* of 1664, John Evelyn declared that he was 'no great Friend' of pollarding because 'it makes so many *Scrags* and *Dwarfes* of many *Trees* which would else be good *Timber*.' However, he accepted that pollards had their uses, and advised that they should be cut every ten or twelve years 'at the beginning of the *Spring* or the end of the *Fall*'. In subsequent editions of *Sylva*, Evelyn modified his initial text, taking account of the views of other contemporary authors. In 1670 he added that 'The Oak will suffer it self to be made a Pollard, that is, to have its head quite cut off' and in 1706 that 'it may be good for Mast, if not too much prun'd, but not for Timber.' An analysis of the different editions of *Sylva* shows that Evelyn had a nuanced and complicated view on pollarding. His opposition was strongest when pollarding destroyed the potential for producing high-quality timber; yet he was able to see that it could lengthen the life of trees, which he thought explained why many ancient trees in churchyards were pollards. He hated the fact that hornbeams were so frequently lopped for firewood, making them deformed and hollow, but recognized that this did not stop the trees from flourishing. He reported that pollarding of oak and ash meant that they grew in circumference but were usually hollow, that black poplars were frequently trimmed in Italy for vine growing and that the elm should not be topped.[14]

Moses Cook was more positive about pollards than John Evelyn and his writings show that pollarding was a widespread and valued practice in the late seventeenth century. He used the phrase 'good pollard' several times for species such as ash and elm and noted that willows and poplar 'are set for Pollard'. He recommended that pollarded poplars, willows, ash, elm and alder should not be left unlopped too long. Cook saw pollards as a profitable resource, giving as an example the high yield of firewood from growing hornbeam at Hampton Court, and argued for the establishment of 'Water-poplar' pollards for wood production. Even as late as the mid-eighteenth century, pollarding was being written about in a positive manner by one of the most influential agricultural writers. William Ellis, a farmer from Little Gaddesden in Hertfordshire, one of the most widely read farming authors, devoted a chapter to the 'Pollard ash' and one to the 'Standard ash' in his book *The Timber-tree Improved*. Ellis considered pollarding to be damaging to the oak, but for other trees such as ash, alder, willow, elm and maple, he supported pollarding and made 'pollard' one of a threefold categorization of tree shapes along with 'standard' and 'stem'. He noted that 'The pollarded willow is of great service in returning a Top at three or four years end.'[15]

But most commentators in the second half of the eighteenth century saw pollarding as a relict practice which was anti-improvement and a threat to potential timber trees. The enormously influential Arthur Young argued against the use of leaves as fodder. He noted that the French used the leaves of poplar and elm to feed sheep and thought that the young leaves of ivy increased the milk of ewes. He reported that the harvesting of shoots of elm, poplar, ash, hornbeam, white hawthorn and beech was common in both Italy and France, that the leaves of oak and chestnut were mixed with other sorts of leaves and that this forage was seen as excellent. But Young argued against the practice, stating that the feeding of sheep with elm leaves was obsolete. William Marshall asserted in 1796 'We declare ourselves enemies to Pollards' in a phrase echoing John Evelyn's declaration of 1664. Marshall disliked woody hedgerows around arable fields and condemned the presence of pollards, especially low ones, in hedgerows which were the 'bane of corn fields'. He argued that low pollards formed a barrier to air and affected the growth of crops. He also condemned the injudicious lopping of hedgerow trees, but did recognize that hedge pollards provided a valuable supply of fuel and stakes for farmers and were necessary when woodlands and coppices were scarce.[16]

Between 1793 and 1813 the Board of Agriculture commissioned *General Views* to provide a county-by-county picture of agricultural conditions. Of 120 reports, 40 refer to pollarding, shredding or lopping hedgerow trees; most were highly critical, using negative vocabulary to describe these 'infamous', 'injurious' and 'disgusting' activities. John Middleton compared Middlesex favourably to Suffolk, 'whose hedges are filled with pollards of every age, under perhaps two hundred years, of no value to the tenant, and worth to the landlord only a twentieth, or thirtieth, part of what those identical trees would have been worth, had they been protected from the spoliation of the farmer's axe'. He thought that the 'best remedy for this evil would be for every landlord to cut down all the pollards over the whole of his property' and that 'the longer the pollards stand, the less valuable they will be.' Although most authors of the *General Views* were opposed to pollards, some gave evidence of their importance for wood production. W. T. Pomeroy wrote in 1794 that in Worcestershire the 'hedge-rows are everywhere crowded with elm, and though the present custom of lopping and pollarding must certainly injure their growth, they often produce timber of considerable dimensions'. Such wood was used for firewood, to repair gates and buildings and to make agricultural and other tools. Pollarding and shredding were also seen as ways of controlling hedges and reducing the effects of shading on crops. John Clark reported that in Herefordshire the lopping of hedges was essential for crops: 'The lopping of hedge-row timber, while it hurts the beauty of the country by doing a violence to nature, proves beneficial to all vegetables that are placed immediately under the trees.' Moreover, as the branches were necessary 'to mend the hedges, to abolish the practice would be impossible'.[17]

The authors of the *General Views* are virtually unanimous in blaming tenant farmers rather than landowners for the 'abominable practice' of pollarding. Tenants were likely 'to make every tree a pollard' and to 'prune or lop his landlord's trees, under the mistaken notion, that it improves their growth'. Pollarding was a way for tenants to harvest wood on a regular basis; it could also be a means of affirming a tenant's rights over trees. In Bedfordshire pollarding was seen as the appropriation of trees by farmers: 'this abominable practice generally originates with neglect of the quickset hedges, which when fallen into decay, must be supplied by dead ones' and 'the farmer ascends the neighbouring trees, to lop off the necessary materials'. Moreover to ensure 'a constant supply of hedging stuff, he cuts off the leading branches, and afterwards claims the succeeding crop as his own'. In Radnor it was reported that 'tenants of superior ingenuity' appropriated trees by pollarding. In some cases 'the property in trees, which by their own or their predecessor's negligence had been suffered to arrive at maturity . . . became the property of the landlord.' To get this property back 'a few of the lower boughs are cut off and burned out of sight' by the tenant. The following year 'a few more share the same fate; the third year the top is cut off, after which the tree becomes for ever afterwards the property of the tenant'. By local custom the tenant was then 'entitled to top it as often as he pleases, coming then under the denomination of a *rundle*'.[18]

By the early nineteenth century most agricultural commentators were agreed that pollarding was a practice worth stopping and some castigated it as a type of theft. It was identified as a relict of an earlier agricultural system. Very few commentators discuss the use of leaves for fodder. Those that did, such as James Anderson, who mentions the use of twigs of Scots fir as fodder for cattle and sheep in the cold spring of 1782, saw this as an unusual and exceptional use of leaves rather than a normal practice.[19] The agricultural changes of the seventeenth and eighteenth century had made leaf fodder superfluous for progressive farmers. Hay, lucerne and turnips were used to feed livestock during the winter. In addition the rise of coal consumption in rural areas was by the mid-nineteenth century to make pollarding and shredding much less important as a source of firewood. But how could pollarding be stopped? One answer was for landowners to exert their power through the imposition of new clauses in their leases which stopped their tenants lopping and cutting boughs.

This was the approach recommended by Uvedale Price, who wrote in 1786 'On the Bad Effects of Stripping and Cropping Trees' in the *Annals of Agriculture*, a major periodical venture launched by Arthur Young in 1784 to publicize the views of leading agriculturalists. Price opened his essay by expressing surprise that the management of timber trees on farms was often left to the tenant, 'who is too apt to consider them merely as furnishing him with fuel, and hedge-wood, and to send his workmen to cut off their boughs in what manner they please'. He makes a clear distinction between pollarding and stripping. With the former, 'cutting off the head of the tree causes it to shoot vigorously both at the top, and at the sides'; if such trees were allowed to grow again they can 'swell to a great bulk' and 'make a noble appearance' and even 'produce very valuable timber for purposes that do not require length'. Pollarding, then, could be 'less disfiguring to the country', although 'in one respect' it was 'still more pernicious' than stripping, as it generally affected oaks, which were 'the most valuable of timber'. Thomas Hearne drew just such an overgrown oak pollard at Downton in north Hereford-shire in 1784–6 for Price's friend Richard Payne Knight (illus. 73).

Price thought that 'Stripping a tree to the top (as chiefly practised with elms)' was 'the most pernicious, as well as the most disfiguring' practice. Once stripped, 'the lower part of it shoots out very strongly' but 'the top hardly pushes at all' and if the tree is repeatedly stripped it 'at last decays'. He backed up this argument by reporting a 'very observing timber merchant' who told him that stripped elms, 'being full of holes', were particularly unfit 'for what they are most used in the neighbourhood of London, that is for pipes, as the water is frequently bursting out at the knots unless they are secured by lead'. He went on to argue that although it was 'a very general notion among the common people, that the strip-ping an elm makes it grow faster', this was a misconception derived from the 'shoots being longer and fresher the first year after it is stripped', and that overall the growth of a tree was checked by the removal of boughs as trees need leaves to grow.

To give greater strength to this argument he reported an experiment 'that was made to convince a gentleman of large property, at Ledbury, in Herefordshire, that the custom of stripping elms was extremely hurtful to the timber'. This person

> desired that an elm might be felled that was known to have been stripped to the top twice within a certain number of years, and the particular years when it was stripped exactly remembered. It is well known that trees when sawed across show the increase of each year by circles, and that when a tree grows much in any one year the circle is enlarged, and the contrary when it grows but little; when this elm was felled, the person showed that the year after it was stripped the circle was very contracted, the next year it was wider, and the circles continued to increase with the quantity of boughs till the next stripping, when the circle was again contracted in the same manner.

Price recounted that the 'gentleman was so struck with the truth of this experiment, that from that time he never allowed a tenant to touch any of his trees' and that 'the size and beauty of the elms about Ledbury' were a 'standing proof' of the 'effect this experiment produced in that neighbourhood'.[20]

Price blamed landlords for allowing their tenants to convert timber trees into pollards and hence 'the profit of trees on each farm is, in a manner, transferred from the landlord to the tenant.' He recognized that tenants would object 'that if they are not allowed to crop or strip their trees, they can neither get fuel nor hedge-wood, and that their hedges will be hurt by the trees growing over them'. Moreover he was told by some farmers that their workmen 'made it a point of honour to get as high as possible, and that they despised a workman who left many boughs at top'. Price argued that a compromise would be 'to allow the tenant to take off the lower boughs to a certain height, as one quarter, one third, or at most one half of the height of the whole tree'. Price felt that this could be achieved by modifications to tenancy agreements. In addition, however, he argued that landlords themselves had to be much more active in their management of timber on farmland, to the extent that every tree, whether in a hedge or within a field, should be documented.

His principal recommendation was 'to number all the trees on each farm, and in every piece of ground, and to enter them in a book, distinguishing the sorts, as oak, elm, ash &c. those that have been cropped, and those that are in hedge-rows from those in the open parts.' He thought it would be 'very useful to have each tree measured in the girth, and roughly valued'; this would allow 'the encrease of each tree, both in size and value' to be seen from the time the account was taken. An additional benefit of this level of management was that 'all tenants would be very cautious how they cropped, stripped, or felled any tree without leave, when there was so certain a method of detecting them.'[21] These suggestions were put into practice a few years later by Price's friend the banker John Biddulph of Ledbury Park. In his 'Timber Book' of July 1817 there is a 'Survey and Valuation of all the Timber Trees and young trees growing on the several Estates in the parishes of Ledbury and Donnington'.[22] Every tree on the farms was 'blazed and numbered with white paint'; the totals listed were 'Maiden Oak Trees': 493; 'Pollard Oaks': 320; 'Elm Trees': 619; 'Ash Trees': 181; 'Poplar Trees': 31; and one 'Wich Elm'. The profusion of pollard oaks and elms is remarkable; sales of trees were noted in this book through the 1820s until 1829.

Price encouraged landlords to take minute control of timber trees through documentation and surveillance. The *General Views* generally took a similar line and the use of anti-pollarding clauses within leases was strongly supported. In Buckinghamshire it was noted that 'Upon Lord Chesterfield's estate the practice of lopping has been prohibited by Mr Kent, in his regulation for the management of that estate.' Nathaniel Kent was a leading land agent who had been a strong influence on Uvedale Price. John Middleton suggested that 'a covenant should be entered into by the tenant, not to top any tree . . . under a penalty of five pounds for each offence.' He thought that 'any tree recently pollarded, would be evidence against the tenant; who should be invariably required to pay the penalty.' But others felt that such clauses were ineffective: 'It is true, in all leases there is a clause – that the tenant shall not cut, lop or top any timber' and that 'this looks very pretty in theory – but behold the practice.'[23]

The ancient practices of pollarding and shredding declined in importance in the three centuries from 1600 to 1900. Many arguments were used against pollarding. First, in the late seventeenth century, pollards were perceived as an old-fashioned way to produce wood or to feed animals. In the context of widespread enclosure and the modernization of land management, which led to a gradual differentiation between those concerned with farming and those concerned with tree and woodland management, pollarding and shredding were increasingly marginal. They were seen as practices that created neither high-quality timber nor high-quality fodder. Second, pollarding was a practice that was seen to have few benefits for the rising class of improving farmers and estate owners. It was tainted by association with small farmers and commoners and was a nuisance to estate proprietors. Active pollarding was to every passing member of the gentry a highly visible reminder, scattered along hedgerows and within fields, of the activities and practical demands of small farmers. Third, pollarding was increasingly seen as an unnatural element of a landscape that was being consciously redrawn and celebrated as a naturalized landscape. These three groups of negative arguments transformed pollarding from a relatively benign, bucolic activity in 1600 to one seen as backward, barbaric and even threatening in the late eighteenth century.

The critique of pollarding gained greater force because of the increasing concentration of land ownership, the continued decline in the importance of leaf fodder and the rise in the use of coal to replace firewood as a fuel in country districts. In general terms the authors most critical of pollarding and shredding were also the most influential. Minor authors provide a more practical view of the practice and in some cases espouse pollarding as profitable and beneficial. Certainly by the nineteenth century, pollarding was generally seen as an outdated, exploitative practice that damaged an important resource. But the rights to lop trees were still jealously guarded in particular places until the end of the century, the most famous instance being the battle over such rights in Epping Forest.

By the mid-nineteenth century the management of Epping Forest was in disarray and the growth of London and rising property values brought enormous pressure for the building of new houses. The Lord Warden until 1857 was the notoriously extravagant Lord Mornington, a nephew of the Duke of Wellington, whose obituary in the *Morning Chronicle* claimed that 'he was redeemed by no single virtue, adorned by no single grace.' He was keen to enclose sections of the forest and profit from property sales. Landowners, who were often magistrates, spread the belief that the rights of the poor to lop trees and gather firewood were defunct. Although a minority of landowners defended the open forest landscapes, as they increased the value of their estates, most saw the continuance of such rights as encouraging 'waste' and 'demoralisation', and favoured enclosure and the erasure of forest rights, which were 'primitive estates and interests founded on the rules of antique societies and trammelled by vexatious usages'.

Three Loughton men, Samuel and Alfred Willingale and William Higgins, who obtained their livelihood from pollarding and selling firewood, were found guilty of malicious trespass after lopping trees. They chose to go to Ilford Gaol in 1866, rather than pay a fine and costs, to publicize their claim to ancient rights. These disputes were influenced by more general demands for public recreation promoted by the Commons Preservation Society from 1865 and complicated by the legal disputes and debates associated with easements and *profit à prendre*. Eventually the remaining wastes were 'secured for public use' under the Epping Forest Act of 1878. Queen Victoria, who opened the forest in May 1882, noted in her journal that 'the enthusiasm was very great' and that it had 'been given to the poor of the East End, as a sort of recreation ground'.[24] The terms under which Epping Forest was saved from clearance provided that the woods should remain in a 'natural state'; the new Commission stopped the practice of lopping in order to encourage

54 Doorway at Lopping Hall, Loughton, Essex, 1884; architect Edmond Egan.

the growth of 'ornamental trees'. Compensation was paid to the villagers of Loughton and £6,000, known as the Lopping Endowment, was given for the construction of a large working men's hall in 1884. To this day, Lopping Hall, as it is known, remains a memorial to the lost and ancient practice of pollarding. The Arts and Crafts designer and social campaigner William Morris celebrated the appearance of the hornbeam woods where the lopping had ceased and argued that 'I would leave them all to nature' and not to the villager's axe. But the many remaining uncut and rapidly growing pollards caused management problems from the 1890s onwards and there were extended debates over the future of the unnatural practice of pollarding. The debate was won by the middle classes, backed by 'aesthetic tastes and scientific principles which were national or international rather than local', only to be reversed 80 years later with the reintroduction of pollarding by land managers informed by the research of historical ecologists.[25]

By the 1950s, in general terms, pollarding had become a moribund and almost forgotten practice in Britain. It is only in the last twenty years or so that there has been increased interest in it. This is due largely to the survival of the Picturesque aesthetic and the associated desire to maintain ancient pollards, and the recognition of their importance as a habitat for rare beetles, spiders and fungi. Such conservation pollarding is as yet in its infancy but there is now great enthusiasm, encouraged by organizations such as the Ancient Tree Forum, for the recording and mapping of surviving ancient pollards across Britain and more widely in Europe. Great attention is being given to developing effective and safe ways of managing and conserving large, overgrown pollards and establishing new ones. However, much local knowledge associated with the different stages of the practice – the lopping of branches, the drying of leaves, their storage, the nutritive value of the different species and their influence on animal health – has already been lost.

SIX
Sherwood Forest

THE same trees and woods are perceived very differently by different people at the same time, and by groups of people through time. These different meanings and values add rich layers of association and understanding to trees and woods, which can be examined by exploring changing interpretations of Sherwood Forest. Foresters, landowners, archaeologists, historians, writers and tourists from the eighteenth century to the present have recorded contrasting and conflicting views of the ancient Sherwood oaks. The Royal Forest was established in the twelfth century and in the thirteenth century covered an area about 20 miles long and 8 miles wide on the dry sandstone heaths of Nottinghamshire. At this time the forest was characterized by a shifting mosaic of unenclosed oak and birch woodland and heath with many temporary arable enclosures known as brecks. Several of those oaks which survive today are over 500 years old. These trees have at different times, and sometimes at the same time, been valued for naval timber supply; for their connections with Druidic rites; for their 'essential' Picturesqueness; as symbols of aristocratic power; for their Robin Hood associations; as a habitat for rare insects; and as relict 'native' oak populations.

Romance

Sir Walter Scott (1771–1832) almost single-handedly created the most famous imaginative historical forest in the world in 1820. He did this by setting a crucial meeting between Richard 1 and Robin Hood within Sherwood Forest in his historical romance *Ivanhoe*.[1] This novel was a runaway success in Britain, the United States and around the globe and went into many editions throughout the nineteenth century. Scott was enormously influential in bringing about a revival of interest in the Middle Ages and created at his home Abbotsford in the Scottish Borders a gothicized mansion surrounded by many acres of trees, enthusiastically planted according to Picturesque rules.[2] But by powerfully evoking a medieval world of friars, maidens and battling monarchs he also unwittingly unleashed a desire by many to visit Sherwood Forest.

In the novel Sherwood Forest is only loosely placed somewhere between Ashby-de-la-Zouch and York, but the quality of the ancient trees is precisely drawn. A 'mysterious guide' leads characters to a 'small opening in the forest, in the centre of which grew an oak-tree of enormous magnitude, throwing its twisted branches in every direction. Beneath this tree four or five yeomen lay stretched on the ground, while another, as sentinel, walked to and fro in the moonlight shade.' The meeting between Richard Coeur-de-Lion, who had 'the brilliant, but useless character, of a knight of romance', and Robin Hood took place 'beneath a huge oak-tree' and a 'silvan repast was hastily prepared for the King of England, surrounded by men outlaws to his government, but who now formed his court and his guard. As the flagon went round, the rough foresters soon lost their awe for the presence of Majesty'. In this company 'Richard showed to the greatest imaginable advantage. He was gay, good-humoured, and fond of manhood in every rank of life' and was happy to call 'Robin of Sherwood Forest' the 'King of Outlaws, and Prince of good fellows!'[3]

Sherwood Forest was not without literary connections before *Ivanhoe*. William Shield's comic opera *Robin Hood, or Sherwood Forest*, which opened at Covent Garden on 17 April 1784, was frequently performed throughout the late eighteenth century and *Sherwood Forest or Northern Adventures*, a now obscure novel by Mrs Elizabeth Sarah Villa-Real Gooch, was published in 1804. But it was *Ivanhoe* which catapulted Sherwood Forest into the international popular imagination of the early nineteenth century. Soon tourists demanded to explore in Nottinghamshire the Sherwood Forest so vividly yet imprecisely imagined by Scott. The impact of the novel can be seen through the eyes of the American author Washington Irving, who rode from Lord Byron's former home Newstead Abbey (illus. 55) to the Forest

> among the venerable and classic shades of Sherwood. Here I was delighted to find myself in a genuine wild wood, of primitive and natural growth, so rarely to be met with in this

55 After William Westall, *Newstead Abbey*, 1832, etching and engraving.

thickly peopled and highly *cultivated* country. It reminded
me of the aboriginal forests of my native land. I rode through
natural alleys and greenwood glades . . . What most interested
me, *however,* was to behold around the mighty trunks of veteran
oaks, the patriarchs of Sherwood Forest. They were shattered,
hollow and moss-grown, it is true, and their 'leafy honours' were
nearly departed; but, like mouldering towers they were noble
and Picturesque in their decay, and *gave evidence, even* in their
ruins, of their ancient grandeur.

He relished the literary associations:

As I gazed about me upon these *vestiges* of once 'merry
Sherwood' the picturings of my boyish fancy began to rise in
my mind, and Robin Hood and his men to stand before me
. . . The horn of Robin Hood again seemed to sound through
the forest. I saw his sylvan chivalry, half huntsmen, half free-
booters, trooping across the distant glades, or feasting and
revelling beneath the trees.[4]

This contagious enthusiasm for visiting untrammelled tracts of wild medieval forest precisely coincided with the final stages of the dissolution of the ancient forest laws at Sherwood. Although some of the remnants of the medieval English Royal Forests, such as the New Forest and the Forest of Dean, remained actively managed by the Office of Woods throughout the nineteenth century, most, including Sherwood Forest, were disafforested and sold off to private landowners. Like other medieval Royal Forests, Sherwood had come into being to protect monarchical hunting rights.[5] Most forests contained villages, heaths, arable land, pasture and woodland. A survey of Sherwood Forest made by Richard Bankes in 1609 showed that by far the greater part of the forest was agricultural land and heath and that it even included the whole town of Nottingham.[6] There was no direct connection between the idea of forest and the concept of woodland but with the decline in Crown interest, especially from the eighteenth century onwards, the term 'forest' became increasingly associated with those wooded areas which survived in areas that remained or were once Royal Forests in the legal sense.

In the Victorian period the ancient trees found in many forests became increasingly celebrated by tourists and authors. Trees mirrored human existence; in 1826 Jacob Strutt restated a commonplace when he noted that

> among all the varied production with which Nature has adorned the surface of the other, none awakens our sympathies, or interests our imagination, so powerfully as those venerable trees . . . silent witnesses of the successive generations of man, to whose destiny they bear so touching a resemblance, alike in their budding, their prime and their decay.[7]

Some areas that had many old trees, especially if they were of a grotesque appearance and had legendary connections, such as Sherwood Forest, became important tourist attractions. There was great antiquarian interest in the ancient trees which soon became overwashed by Picturesque sensibilities. These provided fertile ground for the rapid spread of Romantic ideas drawn from *Ivanhoe* and developed by both local authors and the aristocrats who now owned the land. Notwithstanding scholarly research that indicates that the real or mythical figure of Robin Hood had only the loosest of connections with Sherwood Forest, he

remains a dominant cultural association.[8] But this is only one of the
layers of meaning which have become attached to the trees. The same
individual trees, sometimes alive, sometimes dead, have been ascribed
a catalogue of changing values and meanings. They have been prodded
and probed, lopped and pollarded, exploited and felled. They have
designated status and power and caused legal disputes. They have been
the subject of archaeological experiment and aesthetic reflection. They
have been categorized as fuel, timber, Picturesque, dead and wildlife
habitat.[9]

Corruption

In the late eighteenth century the surviving Royal Forests were identified both as anachronistic and a potential source of revenue. A Royal
Commission was established to examine the extent of surviving Crown
rights. The report on Sherwood was published in 1793 and undertaken
by Charles Middleton, John Call and John Fordyce of the Land Revenue
Office, based in Scotland Yard, London. It drew on a wide range of
evidence that included perambulations of the forest preserved in records
kept in the Tower of London and in the Court of Exchequer and

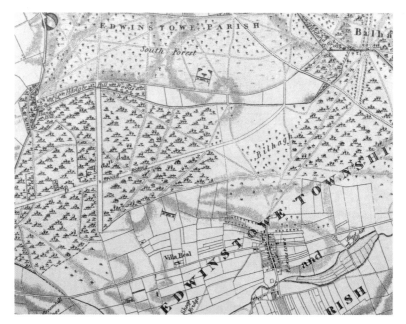

56 George Sanderson, Birkland and Bilhagh, Nottinghamshire, in a detail from an
1835 *Map of the Country Twenty Miles round Mansfield . . ..*

57 The Birklands, *c.* 1890.

various surveys of the forest. They also 'thought it necessary personally to view the Forest, and make Enquiries on the Spot' and interviewed John Gladwin, steward of the courts of the forest, and George Clarke, woodward of the two main surviving woodland areas, Birkland and Billhagh, in November and December 1791. They learnt from Mr Gladwin that 'there was lying in the *Swan Inn*, in *Mansfield*, a voluminous Collection of antient Papers, supposed to relate to this Forest, and locked up in a Chest, which had formerly been kept in the Castle of *Nottingham*' and they also found 'antient Forest Books and Papers in the Possession of different Gentlemen residing in the neighbourhood of the Forest'. The three surveyors found ample evidence of 'old corruption' within the management of the Forest.[10]

Sherwood Forest, first mentioned in 1154, covered most of Nottinghamshire north of the River Trent until 1227, after which the legal boundaries of the forest were restricted to an area roughly 20 miles long and 8 miles wide, largely on sandy and relatively poor soils. The forest was a shifting mosaic of heathland and unenclosed oak and birch woodland and enclosed arable land and woodland. Some of the arable land was in the form of temporary enclosures, or 'brecks'. Most of the land was in private ownership but large areas were subject to common grazing and other common rights until these rights were extinguished by enclosure. By the end of the seventeenth century the area of woodland in Sherwood had reached its lowest point. The report was damning of the forest officials. It argued that 'the Rights of the Crown had been lost Sight of' and were 'very imperfectly known'. The nine forest keepers 'knew very little of their Walks' and 'never acted' other than in 'receiving their trifling salaries'. They found that no deer survived in most of the forest and that the only remaining property of the Crown consisted of

> the Soil in *Birkland* and *Bilhagh*, where neighbouring
> Inhabitants have Rights of Common; of Timber in a State
> of Decay, and exposed to constant Injury; and of Forestrial
> Rights *over* an extensive District, which tend to obstruct
> Improvement, and lessen the Value of private Property,
> without bringing, at present, any Profit to the Crown.[11]

The commissioners examined old surveys of Birklands and Bilhagh and found that 'the Number and Value of the Trees appear, for two centuries, to have been in a State of continual Decrease.' In 1608 there were 49,900 trees of which 23,100 (46 per cent) were 'Tymber Okes' and 26,800 (54 per cent) were 'Okes not Tymber' which were also classed as 'decayed'. By 1680 the number of trees had fallen to 33,996, of which only 1,400 (4 per cent) were suitable for 'His Majesty's Shipping' and 32,596 (96 per cent) were for the 'Country's use'. Many of the trees were 'frow' and 'shaken' and had many 'dead Knotts'. In 1788/9 George Clarke's survey found only 10,117 oak trees left, of which only 1,368 (14 per cent) contained 'Timber fit for the navy'. The surveyors were keen to establish the value of the surviving trees. Their earlier reports on the Forest of Dean and the New Forest had shown 'how loose the Management in Forests has been' and they were not surprised to find 'many abuses' at Sherwood, which was so far from any dockyards.[12]

But they were appalled to find timber trees of such potential importance for the Navy 'going fast to Decay' and made special enquiries. George Clarke confirmed there were about 2,000 acres of Crown trees at Birkland and Bilhagh and that 'no Timber or Wood is cut there for the Use of any private Person.' But when they visited the woods they were shocked by the great age of the trees, most of which 'are now in a State of Decay, and it is not easy to find such as have not some Defect in the Heart, where Trees first begin to fail. This Difficulty gives rise to the greatest Abuse which we have found to prevail in this Forest.' The problem was that various officials could take 'fee trees' as payment for their posts. They could choose the most valuable trees, and to ensure they obtained good ones it was 'a common Practice to bore the Trees first, to see if they are sound; and if a Tree after being bored is not liked, other Trees are tried in the same Manner, until the Party finds one he approves of'. Moreover, the very officers whose duty it was to prevent abuses frequently sold their fee trees to individuals who were able to choose which tree they wanted. The purchaser 'to guard against the Danger of buying [a tree] that is unsound, bores the best-looking Trees to the Heart with an Auger, rejecting everyone in which there is any Mark of Decay'. As this happened every year, many of the better trees were damaged and the surveyors found that 'each Time that we viewed this Forest, we found some which had been recently bored.'[13]

In addition to this extraordinary damage the commissioners found a system of enumerating the trees which itself damaged the trees. George Clarke reported that in 1775 he had been directed to mark and number the trees by 'cutting off a Piece of the Bark about Five Inches Square, and stamping the Crown, the Number of the Tree, and the Name of the Forest, on each Tree, with an Iron Instrument, on the solid Wood. Many of which Numbers and Marks are now partly grown over by the Bark.' The order was changed 'from an Apprehension the Trees might be injured from that Mode of marking', but this was much too late to save the majority from this mode, which had been introduced to help preserve them. The commissioners also found historical evidence of damage to 8,060 trees in Bilhagh caused by the frequent lopping of branches in the survey of 1680. Although the residents of Edwinstowe retained a right to collect firewood in the late eighteenth century, it appears that pollarding had been largely stamped out by that time. In addition to the decline in the number of trees, there was a lack of regeneration of young oak trees, probably due to grazing. George Clarke told the commissioners that the parishioners of Edwinstowe 'claim a Right to the Acorns, when they fall, and take in Swine to feed on them, at certain Rates *per* Head, according to the Plenty or Scarcity of Food. They also depasture their Sheep in those Woods.'[14]

The corruption uncovered by the commissioners was widespread. Forest officers claimed excessive expenses and officers were appointed who 'for many Years, have been chosen for no other reason than to entitle them to their Fees'. The officers were destroying the very trees they were paid to protect because the trees themselves were used as a medium of exchange. And to gain the maximum value of a fee tree, it was bored to the centre to check it was not hollow. The remoteness of the Forest from naval dockyards, together with the enormous power of local landowners who were the principal forest officers, had resulted in a massive decline in the power of the Crown in this remnant of Sherwood Forest. It was only after a thorough examination of surviving documents that the commissioners were able to demonstrate that the Duke of Portland's claim to own the soil of Birkland and Bilhagh was unfounded in law. And yet it was to the Duke of Portland that the commissioners recommended the land should be sold. The report emphasized that it was under Crown ownership and control that the abuses it uncovered had been allowed to unfold and develop. It saw private ownership as the only satisfactory way of encouraging the future management of Birkland and Bilhagh.

Privatization

The rump of Sherwood Forest in the 1790s was an isolated island of Crown land in a sea of private property. Aristocratic landownership was so dominant that the northern part of the forest was nicknamed 'the Dukeries'. These landowners produced a landscape of 'improvement' consisting of parks, mansions, plantations and modern agriculture associated with the great estates of Clumber, Welbeck, Thoresby, Rufford and Worksop Manor which were wrought from the poor sandy soils of Sherwood Forest and from smaller landowners. One expression of the dominance of these great landowners was the construction of long, wide, straight rides through the old woods. One was cut through Birkland and Bilhagh in 1703, and in 1706 another was made from Thoresby House through Bilhagh Wood. The Commissioners of 1793 noted acerbically that in 1709 the Duke of Newcastle (Lord Warden of the Forest) had cut 'a very broad Riding . . . through the Whole of *Birkland Wood*, from one End to the other; and the Timber, which was valued at £1,500 was given to his Grace; but the Expences attending the Fall, amounting to £118. 17s. 2d. were charged to the Crown.' These rides allowed neighbouring ducal owners to take full advantage of the Crown Forest for hunting: riders could traverse the woodland quickly and view the deer clearly. In the case of the Duke of Newcastle, his local power was so great that he was able to charge the Crown for the cost of making the ride and take the profit from the trees felled. The rides demonstrated to all and sundry the status and power of the landowners.

They were linked with avenues of trees, which radiated out across the open agricultural lands in complex geometrical patterns advertising a controlled and subordinated landscape.[15]

Towards the end of the eighteenth century, the old Sherwood oaks attracted the attention of Major Hayman Rooke, a retired army officer who lived at Woodhouse Place just outside Mansfield. He has been described as 'the real pioneer of archaeology in Nottinghamshire' and was a fellow of the Society of Antiquaries. A man of wide interests he excavated a Roman Villa at Mansfield Woodhouse and published a meteorological register from 1785 to 1805. Humphry Repton in his memoirs vividly describes in a scene reminiscent of *Northanger Abbey* how Rooke came to his aid when Repton thought he had discovered a corpse in the middle of the night in a remote room at Hardwick Hall. Towards the end of the eighteenth century Rooke turned his attention to the ancient oaks of Sherwood.[16] He allows us to view the Sherwood oaks through the eyes of a scholarly, late eighteenth-century antiquarian. The trees are celebrated for a variety of reasons. There is wonder at their size, form and dimensions; there is keen interest in their age; there is fascination with their royal and other historical associations. These aspects are combined with a desire to dissect the trees with a scientific purpose: to gain knowledge of their origin and age.

Rooke was keen to marshal statistics about the decline in the number of trees and the scale of destruction and drew heavily on the Report of the Commissioners of 1793. But he combined this statistical approach with an attempt to link the oaks with both the classical world and the ancient Britons and more particularly the Druids. He thought that the

> venerable and majestic Oak seems to claim superiority *over* all other trees. It was styled by the ancients *Jovis Arbor*; and the Celtic statue of Jupiter was a tall oak. Our ancestors, the ancient Britons, held the oak sacred; and their priests the Druids, who took their name from the British Derw, an oak, and esteemed the mistletoe of that tree *above* that of all others, consecrated *groves* of oaks as one species of temple worthy of their religious ceremonies.

He then connects this historical understanding of the ancient trees with the experience of contemporaries visiting the oaks: 'Were we, even now, to enter a grove of stately oaks, seven or eight hundred years old, whose spreading branches form a solemn and gloomy umbrage, I think we could not behold them without some degree of veneration.'[17]

Rooke drew on classical and modern authorities to ascertain the possible ages of oaks. He argued that it 'has generally been thought, that the age of an oak seldom exceeds three hundred years', but that this is 'certainly an erroneous calculation'. He used archaeological and antiquarian knowledge to attempt to date the trees. He considered the principle of the annual growth rings of trees and quotes from John Evelyn:

> It is said, that the trunk or bough of a tree being cut transversely, plain and smooth, sheweth several circles or rings, more or less orbicular, according to the external figure, in some parallel proportion one without the other, from the centre of the wood to the inside of the bark, dividing the whole into so many circular spaces . . . It is commonly, and *very* probably, asserted, that a tree gains a new ring *every* year.[18]

But Rooke is able to go beyond paraphrasing John Evelyn. Using his own local knowledge and observation he recounted how:

> There are now and then opportunities of knowing the ages of oaks almost to a certainty. In cutting down some trees in Birchland . . . letters have been found cut or stamped in the body of the tree, marking the king's reign, several of which I have in my possession. One piece of wood marked J.R. [Jacobus Rex] was given me by the woodman, who cut the tree down in the year 1786.

The woodman told him that 'the tree was perfectly sound, and had not arrived to its highest perfection. It was about 12 feet in circumference.' Rooke was told that 'the letters appeared to be a little *above* a foot within the tree, and about one foot from the centre; so that this oak must have been near six feet in circumference when the letters were cut.' Rooke went on to argue that a 'tree of that size is judged to be about one hundred and twenty years growth. If we suppose the letters to be cut about the middle of James the First's reign, it is 172 years to the year 1786, which, added to 120, makes the tree 292 years old when it was cut down.'[19]

58 Hayman Rooke, *Plantations at Welbeck*, 1790s, watercolour. Hundreds of acres of broadleaved trees were planted on private estates in Sherwood Forest. Rooke depicts the browse-line caused by deer.

Rooke was a detective who dissected the trees in order to understand their origin; he was also a publicist who espoused the aristocratic Whiggish landed interest. He organized a special celebration in 1788 of the centenary of the Glorious Revolution. He identified even the old and decrepit Sherwood oaks with the greatness of the British Navy and elided their Druidic ancestry with the new plantations of the aristocracy, relishing with pleasure the efforts being made 'to adorn this ancient Forest in a manner truly patriotic and worthy of imitation; the many respectable Persons, whose Mansions and Parks border on the Forest, *have* made, and continue to make, large Plantations in honour of the splendid Victories gained by our gallant Admirals.'[20] His views are those of the establishment, and were heavily influenced by John Evelyn, whose *Sylva* was reissued in many editions in the eighteenth century and remained a staple of libraries of the landed gentry. The spirit of improvement backed by the Industrial Revolution and by the demands of modern naval warfare clothed the heaths of Sherwood Forest with new private, rather than Crown, plantations of pine and oak.

Tourism

From the 1820s onwards, tourists began to visit Sherwood in pursuit of Walter Scott's Robin Hood and the forest they so wanted him to have known. But the ancient oaks had been disappearing fast, the heaths were rapidly being converted to modern agriculture and the medieval parks of Bestwood and Clipstone were enclosed. The vibrant medieval identity created by the popularity of *Ivanhoe* began to exert a protective shield over the remnants of the forest. Local authors were taken by the medieval romance of the forest and wrote poems and essays in celebration of it. Aristocratic owners incorporated the chivalric and medieval motifs into their new estate buildings and supported the development of tourism by maintaining the surviving ancient trees and allowing organized parties to visit the forest. The trees also fitted perfectly with the Picturesque ideals of landscape appreciation that William Gilpin and Uvedale Price had presented in the late eighteenth century and which soon became the dominant arboreal aesthetic. The ancient Sherwood oaks, however hollow and rotten, found themselves at the forefront of fashion.

One of the most effective local proselytizers for Sherwood was William Howitt (1792–1879), whose father was a mine superintendent and whose mother was a herbalist. He was born in Derbyshire and apprenticed to a cabinetmaker in Mansfield in the heart of Sherwood. Enthused by the forest landscape, he started to write essays and poems influenced by Byron, Scott and Washington Irving. He became a pharmacist in Nottingham and around him and his wife Mary, who published *The Forest Minstrel* in 1823, gathered a group of poets and authors known as the Sherwood Group. These included William's brother Richard Howitt (1799–1869), called 'The Wordsworth of Sherwood Forest'; Robert Millhouse, who published *Sherwood Forest and Other Poems* in 1827; and Spencer Hall, the phrenologist and poet who published the *Sherwood Magazine*. William Howitt's essay on Sherwood Forest in his influential collection *The Rural Life of England* (1838) marvelled at the great age of the trees:

> A thousand years, ten thousand tempests, lightnings, winds, and wintry violence, have all flung their utmost force on these trees, and there they stand, trunk after trunk, scathed, hollow, grey, gnarled; stretching out their bare sturdy arms, or their mingled foliage and ruin – a life in death . . . it is like a fragment of a world worn out and forsaken.

He contrasts this Sublime yet pantheistic vision with the life of the town worker and celebrates their potential for recreation. 'These woods and their fairy-land dreams are but our luxuries; snatches of beauty and peace, caught as we go along the dusty path of duty. The town has engulphed us; a human hum is in our ears; and the thoughts and cares of life are upon us once more.' At the same time he closely notices the work of the woodmen who 'felled trees that were overtopped and ruined by their fellows; and their billets and fallen trunks, and split-up piles of blocks, are lying about in pictorial simplicity'.[21]

A fascinating insight into the life of the workmen of Sherwood is provided by Christopher Thomson, who published his *Autobiography of an Artisan* in 1847. Thomson settled in the Sherwood Forest village of Edwinstowe as a painter and decorator. The author January Searle found him working in his garden 'and directly he saw us enter the gate, he dashed his spade in the ground, and came forward with his hearty right hand, to welcome us'. He established an artisan's library and reading room and was 'a landscape painter, too, and his pictures of forest scenery are as truthful as Nature herself. We found the walls of his parlour covered all over with the works of his easel. He is married with twelve or thirteen children – a whole house full at least – although I am not sure about the exact number.'[22] Thomson's artisanal Picturesque views, together with his written descriptions, provide a remarkable entry into mid-Victorian rural sensibility (illus. 59, 60).

His writings nostalgically recall the loss of common rights to fuel and fodder following the sale of the Crown lands:

> Half a century ago, the people of this district retained many
> of the privileges which anciently belonged in common to the
> inhabitants of Sherwood Forest. Here the hays of Birkland
> and Bilhagh . . . were open for them to range in, and indirectly
> to profit by; . . . they could supply themselves plentifully with
> firewood, during the whole year.

In the autumn the women and children harvested the bracken, burnt it and sold the ashes to alkali manufacturers. This produced enough income 'to pay off the year's shoe bill for the family, or . . . other tradesman's bills'. They could also 'turn their swine into the forest, where they were allowed to fatten upon the mast'. In addition there 'was no lack of amusement' as 'the villagers residing on the forest skirts, could go forth with their guns, and kill the young jackdaws, starlings or small birds, without having fear of gamekeepers and trespass warrants before their eyes.'[23]

59 Christopher Thomson, 'The Major Oak, Sherwood Forest' (top).

60 '"Simon the Forester", Sherwood Forest' (bottom), from *The Hallamshire Scrap Book: containing views of Hallamshire, Derbyshire, Notts., and adjoining counties. Drawn on stone from nature* (c. 1867).

All this had changed when the woods were privatized; but the new owners had made some improvements. Thomson's view of the Edwinstowe of 1857 'as it appeared ten years ago' shows a row of cottages: 'as far as broken lines go they are picturesque indeed. But . . . cold clay floors, want of room . . . a chamber, in which a family of five persons and upwards all lay down'. His accompanying text points out that 'some would doubtless prefer the group of modern cottages erected by the good Earl Manvers for the use of his workmen, where comfort and cleanliness have each a local habitation, and where useful garden plots worthy of the name are added.' Thomson also shows how trees became renamed to make them fit for tourists. He describes how 50 years ago the 'Cockhen Tree' had 'reigned unobserved in the centre of a ring of birches . . . his locality known only to a few . . . an old farmer . . . [kept] game fowls in the wasting heart' of the tree with 'an improvised oaken door of rude construction.' However, when

> the vast size of this noble tree began to excite the curiosity of the outer world one by one the birches around him began to fall before the axe, and the noble Earl . . . instructed his chief Forester to give him fair play – banish the game cocks – and let the old monarch look royally around on all he could survey. He was locally re-christened and called the 'Major Oak,' in compliment to Major Rooke, the celebrated antiquarian.[24]

Tourist guides such as James Carter's *A Visit to Sherwood Forest including the Abbeys of Newstead, Rufford, and Welbeck . . . With a Critical Essay on the Life and Times of Robin Hood* (1850), began to appear. The great landowners were quick to appropriate this democratic symbol. The Duke of Portland built a new lodge at Clipstone in the forest in 1844 (illus. 61). The main room over the arch was 'dedicated by its noble founder to the cause of education, for the benefit of the villagers of Clipstone'. 'The prospects from this room are most beautiful, including Birkland with its thousand aged oaks, the venerable church of Edwinstowe, and a wide expanse of forest scenery.' January Searle noted 'on the north side, there are statues of King Richard the lion-hearted, Allan o' Dale and Friar Tuck; on the south side there are similar sculptures of Robin Hood, Little John and Maid Marian.' Another guide describes three of the sculptures as 'the ancient frequenters of the neighbourhood: one its presiding deity, Robin Hood; the other Little John; and,

61 The Archway School, Clipstone, *c.* 1890.

bearing them pleasing company, as was her wont formerly, Maid Marian'.[25] These locally inspired buildings were joined by exotic ones such as the Duke of Portland's 'Russian Cabin'; the Russian writer Ivan Turgenev saw the 'semi-feudal, semilibertarian great estates in the English Dukeries' as a possible model for Russian estates after the emancipation of the serfs.[26] The great new mansion built from 1864 to 1875 at Thoresby for Earl Manvers, designed by Anthony Salvin, incorporated a vast library fireplace celebrating Sherwood Forest. Its iconography confirmed the historical connection between Robin Hood and the ancient oaks for the mid-Victorian mind. The huge chimney piece

> consists of an elaborately carved representation, in Birkland oak, of a scene in Sherwood Forest, in which are introduced the venerable 'Major oak' with his knotted and gnarled branches, a foreground of botanical specimens, and a herd of deer – all chiselled with much similitude to Nature . . . Statuettes of Robin Hood and Little John support each side of the piece.[27]

The combination of medieval legend and trees old enough to have witnessed the scenes depicted by Walter Scott was enormously potent. By the mid-nineteenth century the ancient oaks of Sherwood had become firmly fixed in the popular imagination as medieval icons. Within a few years several more individual trees, such as 'Robin Hood's Larder',

'Simon the Forester' and even the 'Ruysdael Oak', were imaginatively named, gaining credence through being identified on Ordnance Survey maps (illus. 63, 64).

The second half of the nineteenth century saw increasing interest in the natural history of Sherwood Forest. Some locals became professional collectors. John Trueman of Edwinstowe was a 'first-rate entymologist, who although a shoemaker by trade, corresponds with the first men and societies in the kingdom, and is known as a valuable entymological contributor to the cabinets of our national institutions'. He told January Searle 'of the haunts, nature, habits, and metamorphosis, of the various insects and butterflies' he collected and how he 'got together one of the rarest cabinets in the kingdom'. On

> dark nights he goes out into the Forest with a pot of rum and honey which he smears over the bark of the trees, to lure the insects he wishes to take. After waiting some time, he pulls a dark lanthern from his pocket, and throws the light full upon the tree, where he beholds his victims enjoying their death-supper, with no small satisfaction. He then quietly brushes them into a tin box . . . and kills them with spirits of camphor.[28]

Many of the most zealous collectors were clergymen and John Carr described the 'rich district of Sherwood Forest' as 'one of the best collecting grounds in the country'. Many 'rare species' had been discovered

62 The Russian Cabin, Sherwood, *c.* 1890.

63 The Major Oak, *c*. 1890.

64 The Ruysdael Oak,
Welbeck, *c*. 1890.

by workers such as the Revd Alfred Thornley, 'who has devoted many years to the investigation of the Coleoptera of the county'.[29] Another enthusiast was the Revd A. Matthews, who found rare beetles associated with the ancient oaks 'by sweeping under oaks' and 'taken in faggots'.[30] The forest also became famous for fungus: over four days in September 1897 the British Mycological Society added 250 species to the known fungus flora of the district. The ornithologist Joseph Whitaker celebrated 'the considerable area of ancient woodland, largely consisting of fine old oaks, with a sprinkling of birch and an undergrowth of bracken' surviving in Sherwood Forest for birds.[31] The ancient oaks were ideal for woodpeckers and in the 1860s an ornithologist pointed out that 'nearly all the old oaks of the forest have suffered the loss of their tops by the agency of wind and lightning, aided by natural decay. Sometimes you see the upper portion of one of these venerable trunks quite denuded of its bark, and riven with many fissures, though the tree is all the while in vigorous growth.' He enthusiastically described how he had

> often noticed the green woodpecker practise a singular feat.
> Placing its bill in one of the long cracks I have mentioned,
> it produces, by an exceedingly rapid vibratory motion, a loud
> crashing noise, as if the tree was violently rent from top to
> bottom . . . It would eventually rouse up all the insects, for
> it seemed as if the tree quivered from top to bottom.[32]

By the end of the nineteenth century Sherwood Forest was a popular tourist destination. Special tours were established around the parks and through the ancient woods. A guide of 1888 extolled the 'wild rusticity of Nature in her primeaval loveliness' to be found in the Forest combined with the 'fertile pastures' of improved agriculture.[33] Earl Manvers, who owned most of Birklands and Bilhagh, commissioned the leading forestry expert W. H. Whellens in 1914 to write a plan for the management of the 1,270 acres of ancient oak and birch woodlands on his estate. Whellens argued that the Picturesque and ornamental value of these woods equalled their timber and game value and that every attempt should be made to 'preserve the natural appearance of these woodlands' for their ancient oaks and the 'Birch with their silvery barks and graceful foliage'.[34] A tourist guide celebrated the 'Assiduous care' that was 'ever manifested by the noble owners that Sherwood, though sadly shorn of her fair proportions, may yet be preserved in Birkland

and Bilhagh in all her primitive beauty and sylvan splendour'. The pleasures to be found were many. The guide emphasized the 'scenes of jollity and mirth within the shadows of the "Major" Oak' and how a 'dozen can enter the hollow trunk at once, and if it is a very gleeful party of youthful Tourists, the girls, on emerging from the cavern, have been known to be subjected to a kiss from the more ungovernable members of the company.'[35]

Destruction

The impact of the massive social and economic changes of the early years of the twentieth century on Sherwood Forest was intensified by exploitation of the underlying coal reserves. In addition to the spread of coal tips over ancient woodland and heath the demand for pit props encouraged the planting of Scots and Corsican pines on the heathland. Joseph Rodgers had noted in 1909 that 'Coal pits are already opened within the near neighbourhood, and no great length of time is likely to elapse before the forest in its beauty, as we know it, will be a thing of the past.' In 1919 adverse economic conditions caused Lord Manvers to lease Thoresby's mineral rights.[36] Income from the colliery's rents, totalling £60,000 by 1931, clearly overcame the pre-war resolve of his agent not to give in to requests for mineral rights on account of protecting the woodland scenery.[37] The unremitting spread of the coal tip did not proceed without resistance: in the 1920s the Beech Avenue, consisting of four rows of trees, which in its time rivalled the Major Oak as a place of local and national interest, was threatened by a proposed colliery railway.[38] The strength of national concern for the avenue, hailed in *The Times* on 16 January 1925 as 'Probably the most remarkable and beautiful woodland sight in England', clearly aided the success of a Manvers-led petition to protect it.[39] During the 1920s and '30s the Forestry Commission, established in 1919, started to acquire large areas of Sherwood Forest heathland from the established estates to make extensive plantations of Scots and Corsican pine. Woodland grants were introduced by the Forestry Commission as an ad hoc measure in 1922 and were later fixed at £2–£4 per acre in 1927 to encourage planting by private landowners.[40]

The impact of the Second World War on the ancient oaks was to be decisive. Military requisitioning began in March 1942 and large areas of the Birklands and Bilhagh ancient oaks were used as stores for

65 Paliama monumental olive tree, approximately 3,000 years old, Paliama, southern Crete, 2011.

66 Japanese alpine forest, Hayachine Mountain, Honshu, Japan, 2013.

67 Thomas Jones, *On the Banks of the Lake of Nemi*, 1777, watercolour drawing.

68 Thomas Bewick, *Red Deer Stag*, *c.* 1780–90, hand-coloured wood engraving.

69 Albrecht Dürer,
Spruce, *c.* 1497, watercolour and body-
colour.

70 Salvator Rosa,
*Mercury and the
Dishonest Woodsman*,
c. 1663, oil on canvas.

71 Giovanni Bellini, *Madonna of the Meadow, c.* 1500, oil and egg tempera on panel. The trees in the background have had their branches cropped for fodder.

72 John Dunstall, *Pollard Oak near West Hampnett Place, Chichester, c.* 1660, watercolour drawing.

73 Thomas Hearne, *Oak Trees, Downton*, 1784–6, ink, watercolour and wash drawing.

74 *Epping Forest,* London & North Eastern Railway poster (designed by
F. Gregory Brown), 1923.

75 *Birnam Wood*, 1801, hand-coloured aquatint with etching by James Mérigot after Hugh William 'Grecian' Williams.

76 Giovanni Bellini, *Assassination of St Peter the Martyr*, 1507, oil and tempera on wood. In the wood four men are felling trees. This is an early represention of coppicing.

77 Invasive *Robinia pseudoacacia* (black locust) near Tokyo, 2013.

78 Paul Nash, *We Are Making a New World*, 1918, oil on canvas.

ammunition; the old trees were now valued as camouflage.[41] In addition large parts of the forest were used for military practice, the sandy heath providing an almost perfect location for tank training in preparation for the north German plain.[42] Coal production increased rapidly and Lord Manvers expressed concern that his young pine plantations had to be felled prematurely for pit props and to allow expansion of the spoil tips. A government official thanked him for continuing 'to sacrifice your young plantations for pitwood. This destruction of immature woodlands is a miserable business, but I am afraid there is no alternative.'[43] The timber merchants argued that 'In sacrificing his woods, he is rendering a very real service to the state.'[44] Even during the war some trees were retained because of their aesthetic value and the Beech Avenue was exempted from wartime clearance in April 1942.[45] But the Earl demonstrated considerably less regard for the ancient oaks than his predecessors. In June 1942, at the same time that he was bemoaning the loss of Thoresby's young plantations, he felt able to offer the Timber Supply Department his old oaks on a plate: 'As regards sawtimber, I have a large number of "stag-headed" oak-trees aged I believe about 800 years, which could be promptly made available if and when required.'[46] The Earl clearly felt that the Picturesque form of the oaks provided them no protection under conditions of total war (illus. 79).

It might be thought that the end of the Second World War would halt the relentless pressure on the ancient Sherwood oaks, but instead it was during the 1950s and '60s that they almost reached their nemesis. Commercial forestry received a boost after the war with substantial tax concessions and government grants designed to encourage the rapid replanting of thousands of acres of woods and the afforestation of heaths and moors. This post-war confidence in commercial forestry was backed by the Forestry Commission's Dedication Scheme for Private Woodlands of 1946, which provided financial assistance for woodlands managed in a state-approved manner.[47] This facilitated the further clearance of ancient trees from 1948 onwards.[48] The fate of many ancient trees was sealed by the arrival of a new Thoresby land agent in 1955 who instigated an extensive planting programme of 60 acres per annum that was to prevail for twenty years and transform many of the estate's woodlands. Planting shifted to the modern commercial system of even-aged, single-species stands. Many new coniferous plantations were established and Corsican pine, rather than Scots pine, became recognized as the optimum species for the sandy soils.[49] The estate was well known for its excellence

79 The Wounded Giant, Sherwood Forest, 1908, photograph.

in forestry practice and was awarded a silver medal for the best managed woods over 500 acres in a competition run by the Royal Forestry Society and the Royal Agricultural Society of England.[50] Many old oaks and young birches were cleared to facilitate the new plantations. One of the estate workers, John Irbe, recollected felling ancient oaks, which were cut into 3-foot sections and split with axes in the 1950s. George Holt recounted how ancient oaks were removed in the 1960s by cutting out square sections to allow the trees to be set alight, their hollow interiors acting like chimneys. Field evidence of this practice remains to this day in some of the surviving ancient oaks.[51]

Other more subtle changes were taking place. The traditional grazing of most of the woodlands was brought to an end through the conversion of pasture and park into arable farmland hastened by farm amalgamation and mechanization.[52] The lack of grazing throughout the forest, together with the destruction of the rabbit population through myxomatosis from 1953 onwards, transformed the vegetation. The extensive areas of bilberry and heather became much reduced and shaded out by the natural regeneration of birch and by the new coniferous trees. In the early years of the twentieth century birch had been highly valued as an essential part of the Picturesque Sherwood ensemble. Grazing stock and rabbits had hindered regeneration, but after the Second World War, when grazing ceased, birch grew so vigorously on the sandy soils disturbed by military manoeuvres that by the end of the century the regrowth was seen as a threat to the characteristic open grazed woodland and heathland. Certain areas of Sherwood were retained for amenity purposes, notably large parts of Birklands, but many were removed in the early stages of the replanting programme or fell into disrepair, as with the Chestnut and Beech Avenues. Chestnut Avenue was fortunate to survive: John Irbe remembered that Lady Manvers accidentally came across men preparing an ancient chestnut to be felled and that this immediately precipitated the sacking of the land agent responsible. But the Beech Avenue was cleared in the years 1976 to 1978 for new plantations following the effects of storm damage, neglect and old age.[53] By the late 1960s many of the ancient oaks had been removed or were rapidly becoming shaded out by the vigorous plantations of Corsican pine or encroached on by the looming colliery spoil.

Conservation

Contrasting views of the ancient oaks in the late 1960s are exemplified by the fate of two sections of Birklands. In one part leased to the Forestry Commission the oaks were simply seen as obstacles in the way of planting conifers and most were felled, while those that remained were rapidly surrounded and shaded by dense pines. In the other part the oaks were the *raison d'être* of a new country park established by Nottinghamshire County Council in 1969. The driving force behind this was increasing demand for public access to popular sites such as the Major Oak, and the lease of the land by the Thoresby estate to the County Council was seen as its 'contribution to the general public'.[54] The council built a large visitor's centre under the ancient oaks including shops, a café, lecture room and other educational and administrative buildings in a 'series of hexagonal huts or "pods" grouped in a compound, inspired by the dugout shelters of early forest dwellers'.[55] Paths were laid out, by far the most popular of which led from the visitor's centre and its extensive car parks to the Major Oak. The role of Robin Hood was maximized and by the end of the century Sherwood Forest had become reconstituted as a tourist destination with over a million visitors a year.

The tide of enthusiasm for lowland commercial coniferous forestry in Britain, especially for plantations on heath and ancient woodland, began to turn in the 1980s. At Sherwood the extent of plantations made over the previous twenty years meant that there was little scope for making new ones. The political consensus on the need for commercial forestry changed in the mid-1970s and many private foresters were increasingly concerned about the implications of additional taxation proposed by the Labour Party. At Thoresby there was a shift in management from planting to consolidation and maintenance of the new plantations. The permanent woodland staff, which in 1984 had consisted of a head forester and nine woodmen, was much reduced by 2000, as with many landed estates, and there was a greater use of contractors.[56] This slackening in

80 Ancient oak at Birklands, Sherwood Forest, 2012.

commercial forestry changed the fortunes of the ancient oaks outside the country park.

Many individual ancient oaks were still to be found in the plantations of conifers; they had survived simply because they had been too difficult or expensive to remove. Moreover significant populations of ancient oaks remained in the country park and other crucially important areas within the old Thoresby Park near Buck Gates. There was growing realization towards the end of the century of the enormous international importance of these trees as habitat for beetles and other insects. The great age of the trees and the quantities of dead wood they contained became valued as an ecological resource. The whole area was designated a Site of Special Scientific Interest and designated under the EU Habitats Directive as a Special Area of Conservation. It was the 'most northerly site selected for old acidophilous oak woods', 'notable for its rich invertebrate fauna, particularly spiders, and for a diverse fungal assemblage' and where there was 'good potential for maintaining the structure and function of the woodland system and a continuity of dead-wood habitats'. Surveys identified the number, condition and distribution of the trees and it was found that there were now about 1,000 veteran trees. A sample of these was bored with holes, rather smaller than those employed in the eighteenth century, to gain cores which could be dated using dendrochronology, thus helping to establish the long-term viability of the living and dead trees.[57]

After over 50 years of neglect, active management by the private estates and the Forestry Commission had returned to the ancient oaks. Conservation agencies provided funding to remove conifers from around the surviving oaks and to fence areas so that traditional grazing could be reintroduced. Cattle were first reintroduced in 1998 when twelve cattle were allowed to graze in the Buck Gates area for six weeks. This experiment indicated that grazing was inhibited by the heavy growth of bracken and so various trials were then undertaken with mechanical rolling and the application of chemical herbicides to reduce the mat of bracken.[58] These experiments proved successful and sheep were introduced to start the conversion of the landscape back to wood pasture. The ancient oakwoods and remnant heathland are now seen as of European importance by ecologists and their future management a matter of more than local interest. The reintroduction of grazing and the removal of coniferous plantations is producing a new landscape, but one very similar to that of the eighteenth and nineteenth centuries.

Moreover, landowners, government agencies and charities such as the Sherwood Forest Trust are working to conserve characteristic fragments of heathland and woodland scattered throughout the ancient boundaries of Sherwood Forest.

SEVEN
Estate Forestry

THE image of Birnam Wood in John Stoddart's description of the local scenery and manners of Scotland published at the start of the nineteenth century shows a wild and ravaged scene (illus. 75).[1] In the foreground is a cut tree stump and beyond this a man is struggling in the strong wind to make his way through a straggling wood; his hat has blown away. The woodland is grazed, and many of the trees have been pollarded; it is a good example of upland wood pasture. The jagged lightning reinforces the Sublimity of the scene, which could be appreciated as a literary and aesthetic landscape. But it would not be one that would be valued by the burgeoning number of landowners whose wealth was increasingly derived from international trade and home industry and who through the nineteenth century strove to plant and manage plantations and woods. The medium-term effect of the agricultural and industrial revolutions was a massive concentration of wealth in the hands of a relatively small number of landowners. Dr Johnson had famously questioned whether 'there was a tree between Edinburgh and the English border older than himself'.[2] But this also implies that there could be, as indeed there were, many young trees. The eighteenth-century spirit of planting, associated with such Scottish aristocratic planters as the Dukes of Atholl, was enthusiastically taken up by landowners throughout Britain in the nineteenth century. Many of the most prominent forestry authors and practitioners were Scottish. And they were largely successful in transforming the pattern of woods on the ground and the distribution of trees within those woods: Shakespeare's moving of Birnam Wood in *Macbeth* prefigured the transformation of British estate woodlands in the nineteenth century.

The tremendous and diverse enthusiasm for trees and woods in the early years of the nineteenth century is captured by Jane Austen in

Mansfield Park (1814). One of the first things that Sir Thomas Bertram, the estate owner, does after a return from his slave plantation in the West Indies is to visit 'his steward and his bailiff; to examine and compute, and, in the intervals of business, to walk into his stables and his gardens, and nearest plantations'. While Sir Thomas, the able businessman who has established the financial basis of the estate, is keen to see the growth of his young plantations, his son Thomas relishes the woods for the game they hold. He tells his father of shooting early in October:

> I went over Mansfield Wood, and Edmund took the copses
> beyond Easton, and we brought home six brace between us,
> and might each have killed six times as many, but we respect
> your pheasants, sir, I assure you, as much as you could desire.
> I do not think you will find your woods by any means worse
> stocked than they were. *I* never saw Mansfield Wood so full
> of pheasants in my life as this year.

Fanny Price, Sir Thomas's niece, celebrates the beauty of trees and their leaves, especially 'The evergreen! How beautiful, how welcome, how wonderful the evergreen!', as a shelter from autumnal winds and speculates on the 'astonishing variety of nature' and that in 'some countries we know the tree that sheds its leaf is the variety, but that does not make it less amazing that the same soil and the same sun should nurture plants differing in the first rule and law of their existence'. When Fanny Price was away from the park in March and April she was saddened 'to lose all the pleasures of spring' and had not 'known before how much the beginnings and progress of vegetation had delighted her', especially 'the opening of leaves of her uncle's plantations, and the glory of his woods'. On her return she is 'everywhere awake to the difference of the country since February'. She saw 'lawns and plantations of the freshest green; and the trees, though not fully clothed, were in that delightful state when farther beauty is known to be at hand, and when, while much is actually given to the sight, more yet remains for the imagination'.

In the summer a small plantation at another estate, Sotherton, was enjoyed for its shade by most of the family: the wilderness, 'which was a planted wood of about two acres, and though chiefly of larch and laurel, and beech cut down, and though laid out with too much regularity, was darkness and shade, and natural beauty, compared with the bowling-green and the terrace. They all felt the refreshment of it, and for some time could only walk and admire.' Fanny Price's enthusiasm for this wood is tempered by her concern for its 'regularity' which did not fit in well with her Picturesque sensibility. She was more concerned over the owner's intention to remove an unfashionable avenue of old oak trees. She lamented the loss: 'Cut down an avenue! What a pity! Does it not make you think of Cowper? "Ye fallen avenues, once more I mourn your fate unmerited"' and hoped to visit it 'before it is cut down . . . to see the place as it is now, in its old state'.[3] Styles and types of tree planting and attitudes to old, young, indigenous and exotic trees symbolized different fashions in landscaping. And the wide variety of woods and plantations established were patterned and shaped differently for game shooting, fox hunting and timber production.

Ancient and modern

By the end of the seventeenth century Britain was one of the least wooded countries in Europe. In broad terms woodland had fallen to less than 5 per cent of the land area compared to the 10 per cent estimated to have existed at the time of Domesday. Moreover, compared to many European countries, the number of trees which could be classed as native to Britain was very low, with fewer than 30 broadleaved species and as few as five evergreens. The geography of tree species resulted from a complicated mixture of environmental conditions and past human activities. Although of small acreage, the woods found in different parts of the country varied considerably: intensively managed oak coppice grown for tannin in Cornwall and Devon; strips of coppiced alder along brooks and rivers in the Midlands; pollarded oaks and ashes and stripped elms found in many hedgerows; while significant areas of native Scots pine were only to be found in Scotland. Partly as a consequence of the small area of woodland, Britain was largely dependent on imported rather than home-produced timber and wood products. In addition there was very little publicly owned forest apart from some small remnants of Crown forests, such as parts of the Forest of Dean and the New Forest. This lack of government-controlled forestry was important, as it meant that throughout the nineteenth century enthusiasts for continental ideas of scientific forestry had the difficult job of attempting to persuade many hundreds of private landowners of the value of the new approach, rather than work with state forest officials, as in parts of Germany.

Partly in response to the small area of woodland and the relative dearth of native tree species, the eighteenth and nineteenth centuries saw a dramatic increase in interest in the management of trees and woods. Between 1750 and 1850 there was considerable anxiety concerning the decline in British woodland, which had, of course, major implications for national security, but also for emerging industry and manufactures such as iron working which utilized charcoal. Landowners were encouraged to plant trees as part of the improvement of their estates by the Society of Arts, which offered an 'honorary premium' of a gold or silver medal for those who had planted the greatest number of trees or the greatest area of ground during a given year. It was hoped that this would help to supply the Navy, cover commons and waste ground, provide employment for the poor, become a resource for industry and further 'the ornamenting of the nation'.[4] All aspects of trees and woods were the subject of debate and discussion, including whether the choice of trees should be founded upon aesthetic, scientific or economic grounds and whether existing woods should be managed profitably. Attempts were made to define and categorize trees scientifically or as native and exotic and to establish how trees outside woods in hedgerows, parks, gardens and fields should be managed. These questions were enmeshed with social and political considerations such as whether communities had access to trees or could use wood.

This interest in trees was part of the enthusiasm for agricultural and rural improvement intimately connected with the rise of British imperial power, trade, industry and wealth. This enthusiasm was equally manifest in diverse representations of trees and forests in paintings, drawings, poetry and literature. Trees were valued as signifiers of property and wealth, of nature and beauty, of age and senescence. For Uvedale Price trees were essential for the Picturesque improved landscape. Rising 'boldly into the air', in beauty they 'not only far excel everything of inanimate nature', but are 'complete and perfect' in themselves. Trees offered 'infinite variety' in their 'forms, tints . . . light and shade', and the 'quality of intricacy', composed of 'millions of boughs, sprays and leaves, intermixed . . . and crossing each other' in multiple directions. Through their many openings, the eye discovered 'new and infinite combinations', yet this 'labyrinth of intricacy' was no 'unpleasant confusion', but a 'grand whole . . . of innumerable minute and distinct parts'.[5]

John Claudius Loudon was the most influential follower of Uvedale Price. He was born in Lanarkshire in 1783 and educated under Andrew Coventry at the University of Edinburgh while also working as a nurseryman's apprentice. His first landscape gardening commission was in 1803 for Lord Mansfield at Scone Palace, Perth, and the first of his many books was *Observations on the Formation and Management of Useful and Ornamental Plantations*, published in 1804. He argued that trees were 'the most striking objects that adorn the face of inanimate nature'. An individual tree was equally pleasing and fascinating through the 'intricate formation and disposition of its boughs, spray, and leaves, its varied form, beautiful tints, and diversity of light and shade'. This made it 'far surpass every other object' and produced a general effect that was 'simple and grand' despite the 'multiplicity of separate parts'. Wood was the 'greatest ornament on the face of our globe'.[6] Loudon was one of the greatest of the early Victorian encyclopaedists and his eight-volume *Arboretum et Fruticetum Britannicum* of 1838 is a vast digest of data concerning trees. He was strongly influenced by the social reformer Jeremy Bentham and one of the dominant themes of the book is the utilitarian benefits of making plantations of trees: just as cereal crops and edible roots supplied food, so trees were scarcely less essential for providing timber, without which there would be no 'houses and furniture of civilised life, nor the machines of commerce and refinement'.[7]

Trees were a crucial element of the network of hedgerows, shelter belts, plantations and clumps that were employed to reconfigure the British landscape practically and visually through the late eighteenth and nineteenth centuries. It was in the long-term interest of landowners to maintain control over the management of timber trees on their estates. Indeed, the management of woodland and trees increasingly became the domain of the landowner as opposed to the farmer through the eighteenth and nineteenth centuries, although this varied from place to place depending on the exact form of land ownership and the nature of rights over the use of trees. Tree and woodland management became increasingly disassociated from other agricultural practices. Hedgerow trees on farms were often maintained and controlled by the landlord. The wholesale domination of woodlands by landed interests was not, however, established without frequent social protest and theft of wood.[8] The taking of dead wood, particularly for firewood, remained commonplace in the nineteenth century and this may well have contained an element of protest against the loss of earlier rights to such wood. The number of thefts suggests that the 'perpetrators may have enjoyed some degree of community sanction': in 1852 at Shelwick in Herefordshire 'an effigy of a man who informed upon some persons for wood theft . . . was erected in a cottage garden and ritually shot and then burnt'; the informer was less popular than the thief.[9]

Woods and coppices

Landed-estate forestry took two main forms in the nineteenth century: traditional woodland management, principally of coppice or coppice with standards, and plantations. The increasing number of plantations meant that many older woods were enlarged and some were joined together by plantations to create screens and to improve the ability of woods to hold game. This often resulted in a patchwork pattern of old and new woods. For much of the nineteenth century the established old coppice wood-lands continued to be cut regularly, although their profitability depended on trading conditions and the changing markets for wood products. The replacement of firewood and charcoal by coal for domestic and indus-trial energy production brought about massive reductions in two of the principal markets for coppice. Yet the rapid rise in population caused strong growth in markets for a wide variety of woodland products used in industry, agriculture and the home. Many of these markets were localized and there were considerable regional variations in the strength of the different markets.[10]

Woodland managers recognized that many of the coppice wood-lands were of ancient origin. James Main argued in 1839 that both historical and geological evidence suggested that 'the greater part of the continent of Europe, as well as its islands, were at an early period almost entirely covered with wood' and that some tracts of forest had been preserved within the royal forests and private parks, while other tracts 'of natural forest are also in existence, occupying broken or marshy ground, or precipitous slopes inaccessible to the plough'. The land agent J. West distinguished in 1843 between ancient woods, which were mainly coppice or coppice with standards, and modern plantations. In the nineteenth century the use of exotic trees was almost entirely restricted to the making of new plantations: the replanting of ancient woodland with such trees was generally uneconomic and such replacement was not

81 *Hop Pickers*, 1803, stipple engraving and etching by William Dickinson after Henry William Bunbury.

likely to take place while coppicing remained profitable. J. Standish and C. Noble, who ran a large nursery at Sunningdale, argued in 1852 that

> It is often the object of proprietors to remove woods which are composed of the ordinary indigenous trees of the country, and to replace them with others of an exotic and more ornamental character. But the advantages of such existing woods are generally too great to allow their removal.[11]

William Ablett was able to emphasize the profitability of coppice as late as 1880: 'Where copse-wood is cultivated to any considerable extent, it is advantageous so to arrange matters to come on in perpetual rotation, which may be cleared and put to profitable use yearly.'[12] In some areas with special local markets, such as that for hop poles in Kent, Herefordshire and Worcestershire, the demand for poles was so high that it was worth planting up new coppices or restocking old ones. Tom Bright, a Kentish agricultural valuer, noted in 1888 that 'the improvements that have been effected of late years in the management of underwoods in many districts, are patent to the most casual observer.' He emphasized that 'there are some natural underwoods, that are capable of much improvement . . . All vacancies in old woodlands [should be filled up] with chestnut or ash plants, according to the nature of the soil, at distances of not less than 6 feet from each other, but at considerably less than that from old and inferior stubs or stools.'[13]

In 1894 John Nisbet, who had experience in the Indian Forest Service and taught at the West of Scotland Agricultural College, also stressed the regional importance of coppice: 'Oak coppice-woods are often by far the most remunerative form of silvicultural crop, when there is any favourable market near at hand for the disposal of the bark to tanneries.' Mixed coppices of ash, field maple, sycamore, sweet chestnut and hazel were 'also exceedingly remunerative through-out southern England, when near to favourable markets for saplings and poles, such as hop districts'. Alder coppice was similarly 'often a more remunerative form of crop than almost any other on land that is suited for the Alder, but which cannot be conveniently drained to serve higher purposes'.[14] But the decline in coppice prices became marked towards the end of the nineteenth century and only nine years later, in 1905, Nisbet described a completely different and very bleak picture of the state of coppice woodland. He called such woods 'the national form of arboriculture' which 'have for the most part become practically transformed into game coverts . . . Yet the copse-woods and coppices were at one time among the most profitable parts of large estates.'[15]

It was not until the turn of the century, by which time metal implements had almost totally replaced locally produced wooden ones, that the market for coppice finally collapsed in most areas. Many of the characters in Thomas Hardy's *The Woodlanders* (1887) gained their living from coppice trades. In a new introduction to the novel composed in 1912 Hardy wrote: 'in respect of the occupations of the characters, the adoption of iron utensils and implements in agriculture, and the discontinuance of thatched roofs for cottages, have almost extinguished the handicrafts classed formerly as "copsework" and the type of men who engaged in them.' The final collapse of coppicing in the late nineteenth and early twentieth centuries was brought about by the further substitution of wood products by new metal and chemical products and changes in agriculture. This is not to say that the coppice industries did not manage to limp on well into the twentieth century in some localities. On the Eastnor estate in Herefordshire, for example, it was not until the early 1930s that the demand for coppice wood reached such a low point that auction lots did not attract any purchasers.[16]

❧ Plantations

Plantation forestry had become a fashionable branch of British estate management in the eighteenth century and developed its own traditions. It was strongly associated with patriotism and improvement and continued to be a core aspect of rural estate management in the nineteenth century. Its place was assured less by any inherent profitability, although this was always stressed by forestry publicists and professionals, than by a belief in the seemly nature of tree growing and the clear benefits for landscape, game and hunting. George Sinclair (1786–1834), a keen Scottish horticulturalist who was for many years gardener to the Duke of Bedford and later partner in a firm of seedsmen, was a keen exponent of planting as part of the more general improvement of agriculture and the economy. He argued in 1832 that

> Numerous instances might be cited from different parts of the
> kingdom where exposed and sterile lands have, by planting,
> been made capable of producing valuable arable crops and
> the best pasture grasses, and of rearing and fattening stock
> of improved breeds. This, in effect, is adding to the territorial
> extent of the country, to its wealth and strength, by conquest
> over the natural defects of local climate, soil and exposure.

A great advantage, for Sinclair, was that

> Judicious planting and the skilful culture of plantations combine
> national and private interests in an eminent degree; for, besides
> the real or intrinsic value of the timber . . . planting improves
> the general climate of the neighbourhood, the staple of soil . . .,
> affords shelter to livestock, promotes the growth of pasture and
> of corn crops, beautifies the landscape, and thus greatly and
> permanently increases the value of the fee simple of the estate
> and adjoining lands.

So the making of plantations not only counter-intuitively enlarged the amount of available agricultural land, but increased the capital value of property and led to greater agricultural production.[17]

The enthusiasm for tree planting was closely linked with greater power of landowners over large areas of land and their ability to fence, enclose and control the way land was used and managed. It was strongly associated with ideas of improvement and parliamentary enclosure and in Scotland with the harsh removal of people and their traditional farming practices known as the clearances.[18] Planting was encouraged by societies such as the Society of Improvers, founded in Edinburgh in 1723, and the Highland Society of 1784. One of the largest tree nurseries, the Perth Nurseries, was founded in 1767 and covered over 30 acres by 1796. The most prominent planters were the Dukes of Atholl, who became particularly famous for their very extensive larch plantations. A few larch plants were brought to Blair Atholl in the 1730s and successive dukes increasingly recognized their potential for forming plantations. The fourth duke is credited with planting 15,500 acres of trees, and larch was one of his favourite species. He was concerned about the supply of wood for the Navy and felt that larch was a good replacement for oak. Indeed he 'planted increasing amounts annually with pure larch as the supplies from the nurseries became better and extended the elevation of his plantations gradually higher, using Norway spruce . . . in the wetter hollows and Scots pine sparingly.'[19]

Many other landowners were enthusiastic tree planters, but few were so keen on planting pure larch. Most used larch as a nurse mixed with hardwoods such as oak, beech, ash or elm. The intention was that the larch trees would help the hardwoods to grow well and that they would be removed to provide income while the hardwoods were left to grow on. On the Earl of Mansfield's estate at Scone, where Loudon had worked, old arable land was ridged and planted up in alternate lines of hardwood and larch at a distance of between 4 and 6 feet. Most of the larch was removed after thirteen to fifteen years. For one such oak and larch plantation made in 1804/5, the Earl of Mansfield received a gold medal from the Society for the Encouragement of Arts.[20]

The early Atholl plantations 'were systematically carried out' and 'planting plans were made with roads and paths laid out, and the ground divided into units of fifty acres'.[21] This did not necessarily result in the domination of the landscape by rigid, geometrically aligned hedges and walls. The artist and diarist Joseph Farington, for example, was rather taken with the plantations at Atholl on his visit to Scotland in October 1801: 'The scenery as it appears from the road from Dunkeld to Blair is uncommonly fine. The Hills enriched with plantations not formally made but mixing in a natural manner with the rocks, and the banks of the river though artless yet have a sufficiently neat & finished appearance to be a pattern for forming such Parks & what are called Landscape Garden Grounds.' He commented that 'Mr Price & Mr Knight might describe the Tay in this part as an example.'[22] Another keen planter of larch was Thomas Johnes of Hafod, Cardiganshire, who radically transformed the farms and tenancies on his estate, made many new plantations with about five million trees and was encouraged to 'embellish his barren patrimony according to "picturesque" principles'.[23]

But what should be the shape and disposition of the thousands of new plantations? And how should the multitude of new trees fit in with the established woods and hedges? The boundaries and shapes of most existing and ancient woods had not been consciously designed, but by the turn of the nineteenth century the massive concentration of land ownership and the enclosure of the remaining open fields gave the opportunity for owners to reshape the countryside. By 1800 many of the trees planted in the mid-eighteenth century were coming to maturity and owners could make informed judgements about the types of plantation that were most effective.

Uvedale Price argued that great care should be taken in the disposition of plantations and the way in which groups of different types of trees fitted into the landscape. As a general rule he thought that 'it is not enough that trees should be naturalized to the climate, they must

82 *Hafod House, Cardiganshire*, engraving by John C. Varrall after Henry Gastineau.

also be naturalized to the landscape, and mixed and incorporated with the natives.' He was a keen supporter of mixed plantations and argued that a 'patch of foreign trees planted by themselves in the out-skirts of a wood, or in some open corner of it, mix with the natives, much like a group of young Englishmen at an Italian conversazione.' He contrasted this with the situation 'when some plant of foreign growth appears to spring up by accident, and shoots out its beautiful, but less familiar foliage among our natural trees' which 'has the same pleasing effect, as when a beautiful and amiable foreigner has acquired our language and manners so as to converse with the freedom of a native, yet retains enough of original accent and character, to give a peculiar grace and zest to all her words and actions'.[24]

In addition to mixing trees within woods, Price argued that new plantations should be planned carefully so that they added to the 'infinite richness and variety' of the landscape yet seemed 'part of the original design'. In practice, however, he found that many plantations appeared to be planted so as to be 'as distinct as possible from the woods of the country' so none could doubt 'what are the parts which have been improved'. He accused owners of wanting to make a spectacle of their enthusiasm and commitment to planting new trees. Indeed he was critical of those who ignored 'the spontaneous trees of the country' and 'excluded' them as 'too common', choosing instead exotic trees 'of peculiar form and colour' which took the 'place of oak and beech'. He thought that 'whatever trees the *established* woods of the country are composed, the same, I think, should prevail in the *new* plantations, or those two grand principles, harmony, and unity of character, will be destroyed.' He also deplored the common practice of filling 'a vacant space between two woods' with firs and larches as instead of '*connecting* those woods, which should be the object'; such 'harsh and sudden contrasts of form and colour, make these insertions for ever appear like so many awkward pieces of patch-work'.[25]

He noted that trees grown in plantations often had an inferior appearance and compared the 'tameness' of 'poor pinioned trees (whatever their age) of a gentleman's plantation drawn up strait and even together' to the 'animation' of 'old neglected pollards'.[26] Here he articulates a clear conflict between the preference of timber merchants for plantations and woods full of straight and even-aged stems and those of the landscaper, who often preferred woods of mixed ages and species of trees. He notes how 'even large plantations of firs, when they are not the natural and the prevailing trees of the country, have a harsh and heavy look, from their not harmonising with the rest of the landscape.' This was especially true where 'one side of a valley is planted solely with firs, the other with deciduous trees.' This situation was made worse by the tendency to plant trees closely together 'to produce some appearance of wood as soon as possible'. Price noted that owners 'seldom' had 'the resolution to thin them sufficiently' and hence they were 'all drawn up together nearly to the same height . . . no variety, no distinction of form can exist, but the whole is one enormous, unbroken, unvaried mass of black', which he equates with Milton's description of his blindness: 'an universal blank of nature's works'. Even worse was the interior of these dense pine plantations: 'a collection of tall naked poles . . . ; above – one uniform rusty cope, seen through decaying sprays and branches; below – the soil parched and blasted with baleful droppings; hardly a plant or a blade of grass, nothing that can give an idea of life, or vegetation'. Price thought that of 'all dismal scenes' this was the one 'most likely for a man to hang himself', except, he helpfully pointed out, that there was 'rarely a single side branch to which a rope could be fastened'.[27]

Two specialized forms of plantation popularized in the eighteenth century were belts or screens of trees, often around parkland, and clumps of trees within the park. The advantage of plantations of trees as screens had long been recognized but by the early nineteenth century such belts had come to be commonplace and Price complained that they blocked views and impoverished the landscape. He criticized an estate owned by Sir Charles Hastings between Ashby-de-la-Zouch and Measham, which he spotted by its 'long line of plantation' which was along the main road. This plantation was 'an unfortunate one for the traveller, as in a year or two it will completely hide the whole of the distance'. Price thought 'a few of the trees in particular parts' should be removed to allow distant views to 'be shown to great advantage & be form'd into many very pleasing compositions'.[28] Price reserved his most caustic criticism for the parkland clump, which had 'the same effect on the great features of nature, as an excrescence on those of the human face' where 'let there be a wart or a pimple on any prominent feature – no dignity of beauty of countenance can detach the attention from it.' He disliked clumps planted with larch because 'the multitude of their sharp points . . . had much the same degree of resemblance to natural scenery, as one of the old military plans with scattered platoons of spearmen, has to a print after Claude or Poussin', while the 'dark

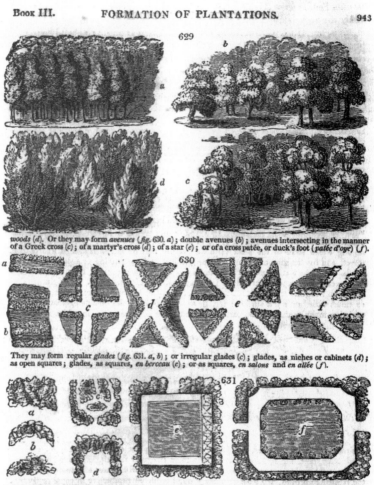

629

woods (d). Or they may form *avenues (fig. 630. a)* ; double avenues *(b)* ; avenues intersecting in the manner of a Greek cross *(c)* ; of a martyr's cross *(d)* ; of a star *(e)* ; or of a cross patée, or duck's foot (*patée d'oye*) (*f*).

630

They may form regular *glades (fig. 631. a, b)* ; or irregular glades *(c)* ; glades, as niches or cabinets *(d)* ; as open squares ; glades, as squares, *en berceau (e)* ; or as squares, *en salons* and *en allée (f).*

631

| 6814. *With respect to the character of tree-plantations,* they may be as various as there are species ; but for general effect and designation, woody plants are classed as large or small, trees or undergrowths, deciduous or evergreen, round-headed or spiry-topped ; and plantations of every form and disposition may be planted with these, either separately or mixed. Thus we have groups of shrubs, groups of high and low growths, and of trees ; plantations of round-headed and spiry-topped trees mixed ; of trees and undergrowths ; or of low growths only, as in copse-woods and osier-plantations.

83 John Claudius Loudon, 'Formation of Plantations', from Loudon's *Encyclopaedia of Gardening* (1830).

tint of' those planted with 'Scotch fir' added the 'last finish' and made them 'horribly conspicuous'.[29]

John Claudius Loudon set out his views on the planting and management of trees in his *Hints on the Formation of Gardens and Pleasure Grounds* (1812), which included plans for laying out gardens and pleasure grounds for large villas and an analysis of plantation management.[30] In his influential *Encyclopaedia of Gardening* (1830) Loudon enthused over the vast variety of types of tree available in his instructions for the formation of plantations, which 'may be as various as there are species; but for general effect and designation, woody plants are classed as large or small, trees or undergrowths, deciduous or evergreen, round-headed or spiry-topped; and plantations of every form and disposition may be planted with these, either separately or mixed.' Avenues might be single or double and might intersect 'in the manner of a Greek cross', 'a martyr's cross', 'a star' or 'a cross *patée*, or duck's foot' while glades might be regular, irregular, 'as niches or cabinets' or '*en berceau*' or '*en salons*'.[31] Many landowners built garden rooms, gazebos or kiosks in the new plantations so that the growing trees could be enjoyed. William Sawrey Gilpin, one of the most prolific landscape gardeners of the mid-century, made plantations which were characterized by 'irregularly shaped plantations with bold projections and recesses'; these mimicked the intricacy 'found in the architectural details of Gilpin's terrace walls with overhanging copings, vases, and protruding buttresses'.[32]

Although Price, Loudon and Gilpin had enormous influence on the mode and form of tree planting in the nineteenth century, on many estates it was the woodman – or, on the larger estate, the forester who had been trained up on several estates – who would make practical decisions about the management and layout of plantations. The planting practices developed by professional foresters on Scottish estates in the eighteenth and early nineteenth centuries were often used on English ones, and by

84 Hayman Rooke, *Turkish Kiosk in Plantation, Farnsfield*, 1790s, watercolour.

the mid-nineteenth century 'it was as much the correct thing for an estate to have a Scotch forester as it was for a nobleman's establishment to possess a French *chef*' and 'the practice of early and heavy thinning, which prevailed in England for fifty or more years, was introduced by Scotch foresters.'[33] William Linnard has identified at least 21 Scottish foresters who worked on the principal Welsh estates, such as Penrhyn Castle, Hafod, Margam and Bodnant, in the nineteenth century.[34]

There was no established school of forestry in Britain and most of those responsible for managing woods and making plantations were practical men who were trained on the estates themselves. Often the long-term planning of forests was in the hands of land agents or owners. The situation was criticized by many commentators and A. C. Forbes, writing in 1904, thought that 'until quite recently (and to some extent even now) it was no uncommon thing to find all classes of men filling the position of estate forester. Any man with a general knowledge of estate work was considered qualified to manage the woods, more especially on those estates on which the area under wood consisted of coppice with standards.' Forbes argued that for such traditional woods 'It required no great ability to manage a squad of half a dozen woodmen, to mark and measure the necessary number of trees for estate use or sale, and to see that hedges and fences were more or less in good condition.'[35] Coppicing thus continued in the tried and trusted way throughout the century.

Forbes argued, however, that the skills required for the successful establishment of plantations were not available on most estates. Indeed, when 'planting was carried on, it was, and still is, usual to get a nurseryman to do it by contract at so much per acre and leave the method of planting and choice of species to him'. He was especially critical of the many small estates where the 'commercial details and higher branches of the work' were in the hands of land agents, who would have only a broad training in forestry, and the 'practical woodcraft' was the responsibility of a foreman woodman, who was 'little more than a skilled workman at the best, with a rule-of-thumb acquaintance with the elements of planting, thinning, draining, and so forth'. The effects of employing such poorly qualified staff were, in his opinion, 'inevitably bad'.[36]

Nurserymen not only provided the trees but wrote many of the manuals for forestry. One of the most influential foresters was James Brown, whose nursery at Craigmill House near Stirling specialized in 'the Coniferous kinds of trees only, all of which he rears from seeds brought from their native localities'. His text *The Forester* (1847) went into many editions and he strove to improve the quality of the timber produced from plantations. He did not hold back from advertising his services:

> James Brown (Author of 'The Forester') begs to inform Landed Proprietors and their Agents, that from having for many years past observed that a large proportion of Larch and Scots Pine plants used in making plantations in Britain are of a description unfit to insure their becoming valuable as timber, he has established a Nursery on a suitable soil and situation in the vicinity of Stirling, in which he rears, on a sound natural principle, plants of the kinds named, and of other Coniferae . . . all of which he rears from seed brought from their native localities.[37]

Brown became famous for continuing to encourage the mixed plantations which were so popular in Scotland. A. C. Forbes, who wrote several forestry texts and lectured on forestry at Durham College of Science, Newcastle upon Tyne, later considered that

> it is to Brown and his school that we owe the introduction of the mixed plantation – a system of planting that has led to some of the worst results that could possibly be attained . . . Pure plantations, when such were planted, invariably consisted of larch . . . The object in many cases was not so much the ultimate production of first-class timber, as the speedy growth of game cover or screens and belts for landscape effect.[38]

But this is unfair, as Brown did a lot to encourage large-scale forestry and careful thinning and management and was at his most insistent when arguing for the careful placement of trees to suit specific localities.

Brown provides a diagrammatic scheme for a mixed plantation (illus. 85). Here hardwoods such as oak, ash, elm and sycamore are each planted 'exactly 20 feet from the next of its own species' while the intervening 'larch nurses' are planted 3½ feet from each hardwood and the 'Scots pine nurses' are 5 feet from the hardwoods. Brown argued that the 'four larches, which are planted next each hardwood plant, can be all taken away, in the way of thinning' before the Scots pines, which need

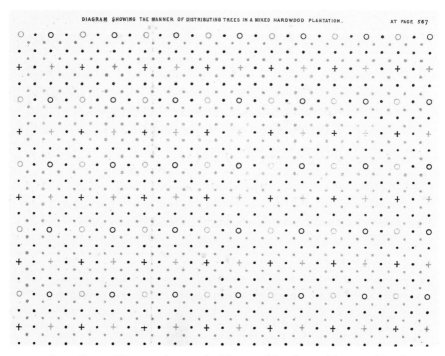

DIAGRAM SHOWING THE MANNER OF DISTRIBUTING TREES IN A MIXED HARDWOOD PLANTATION.   AT PAGE 567

85 James Brown, 'Diagram showing the Manner of distributing Trees in a Mixed Hardwood Plantation' from Brown's *The Forester* (1847).

longer to grow before they 'become of some value'.[39] Brown stressed that every effort had to be made to take into account soil type and altitude when planting and thinning plantations and thought that

> every forester ought to look upon the estate of woodlands
> on which he may have the management, with the eye of a
> geographer. He ought to consider it as a continent in itself;
> each plantation may be looked upon as a separate kingdom
> according to its altitude; and each, again, may, in the mind
> of the forester, be divided into provinces according to aspect
> and latitude, and planted with those trees which are known
> from nature's own rules to be best adapted for it.[40]

If Brown's suggestions were followed closely, then the new plantation landscapes could appear natural and at the same time allow a reserve of timber to become established.

❀ Novel display

Trees were not only aesthetically pleasing, and intellectually inspiring, but through the multiplicity of their agricultural, manufacturing, commercial, building and naval applications they were, according to Loudon, 'the most essential requisite for the accommodation of civilized society'. The pleasure 'attending the formation and management of plantations' was a 'considerable recommendation to every virtuous mind' and young trees could be regarded as akin to offspring; 'nothing' was more satisfying 'than to see them grow and prosper under our care and attention', examine their progress and mark their peculiarities. As they 'advance to perfection', their 'ultimate beauty' is foreseen and a 'most agreeable train' of 'innocent and rational' sensations are excited in the mind, so that they 'might justly rank with the most exquisite of human gratifications'. At the start of his career, in his *Observations* (1804) Loudon was already arguing passionately for the planting, management and improvement of trees as one of the major hallmarks of human civilization.[41]

The popularity and fashionableness of tree collecting made the acquisition, planting and successful establishment of novel tree and shrub specimens, like works of art or antiquities, highly desirable for their own beauty and as a backdrop for parks or as potential timber trees. Nineteenth-century tree enthusiasts were particularly susceptible to the pleasures of evergreens and conifers. The British native flora was bereft of significant evergreen trees other than the broadleaved holly (*Ilex aquifolium*), coniferous yew (*Taxus baccata*) and Scots pine (*Pinus sylvestris*). The box (*Buxus sempervirens*) and juniper (*Juniperus communis*) were also native but were usually grown as shrubs rather than substantial trees. This paucity of evergreens meant that Victorian gardeners prized any introduced trees that could provide varied foliage during the long winter months. One of the greatest enthusiasts for evergreens was the Derby nurseryman and horticulturalist William Barron (1805–1891), who worked for the Earls of Harrington at Elvaston Castle in Derbyshire in the 1830s and '40s before setting up his own nursery business.

Barron's book *The British Winter Garden* promoted the use of evergreens in public and private spaces, helping to drive the new fashion in British, European and American gardens. He was attracted to conifers for their novelty and exotic associations and natural characteristics, which he was able to exploit for economic purposes. Barron and his company helped to foster the mid-Victorian fashion for evergreen planting, promoting them for their economic value and as special and ornamental specimens. Although he propagated and popularized a huge variety of exotic conifers, the evergreen that became most closely associated with Barron was the yew. The ornamental value of the yew and the relative ease with which it could be moved, led him to favour it for many of his landscape gardening and transplanting commissions. He particularly valued the contrasts and effects that could be achieved by placing different forms or colours of yews against different backgrounds, such as the combination of large and small yews behind golden yews, Irish yews, variegated white cedars and different junipers.[42]

For Barron conifers were superior to deciduous trees on the grounds of health, practicality and neatness and provided enjoyment for an entire year. He was critical of the types of deciduous trees that dominated many parks and plantations, complaining how frequent it was to see 'close to our mansions, such commonplace things as elms, ashes, sycamores, poplars, or any other rubbish that the nearest provincial nursery may happen to be overstocked with: all stuck in to produce neither immediate or lasting effect!' Moreover, deciduous trees provided a 'continued litter of decayed leaves', which he thought of as 'an unwholesome effluvia' during the winter, and 'an assemblage of leafless stems' with no

86 Illustration of the grounds of Elvaston Castle, Derbyshire, from E. Adveno Brooke's *The Gardens of England* (1857).

shelter or protection 'from bleak winds for seven months in the year'. In contrast coniferous trees 'excited admiration by providing an infinite variety of form, size, colour, texture and outlines 'from the formal Araucaria and fastigiate Junipers', to the 'wild grandeur of the pine, and even to the delicate, graceful, and flowing habits of the Cryptomeria Japonica… and Hemlock Spruce'. He enthused over gigantic 'Lambert and Bentham pines', *Sequoia sempervirens* and Douglas fir, which towered 'their lofty heads a hundred feet above the pride of British forests'.[43]

Barron's call for conifers did not fall on deaf ears, and he, together with many other nurserymen and authors, helped feed a growing frenzy of enthusiasm for new varieties and species of conifer, which showed itself on the ground in the increasing number of arboreta and pinetums on private estates. The enormous wealth of the Holford family, for example, derived from shares in the New River Company which supplied fresh water to London, allowed them to fulfil their passion for collecting both trees and art. Robert Stayner Holford (1802–1892) displayed his art in vast new houses in London (Dorchester House) and Gloucestershire (Westonbirt), and his rapidly expanding collection of trees in an immense arboretum at Westonbirt. Many of these tree collections were displayed along avenues and rides so that visitors could take in the new varieties. At Bicton, Devon, an avenue of *Araucaria* trees was planted in 1843 from seed sown at the famous Veitch nurseries in Exeter (illus. 87), and one of the surviving trees is now the largest monkey puzzle in the UK, 26 metres tall with a girth of 4 metres.[44] At Eastnor Castle in Herefordshire Earl Somers had a carriage ride 3 miles long 'flanked by evergreen as well as deciduous trees and shrubs', including indigenous and exotic species such as yews, the wild service tree and *Arbutus*.[45]

Covert spectacle

The consolidation of land ownership in the eighteenth and nineteenth centuries was strongly linked with the transformation of vast tracts of land from relatively open areas to enclosed landscapes of largely geometric fields. In very broad terms the former open field zones of the English Midlands were enclosed with hedges, most frequently with hawthorn, and hedgerow trees such as elm and oak were also often planted. In the hill lands of the north and west, the enclosures were more commonly of stone. Although these boundaries are now celebrated as traditional features of importance for nature conservation, for many

87 'The Araucaria Avenue at Bicton', from *Veitch's Manual of Coniferae* (1900).

contemporaries, such as John Clare, they were a potent reminder of the loss of traditional forms of land ownership and farming practices. William Gilpin, journeying towards the Picturesque beauty of the Wye Valley across the Cotswolds in 1770, found that 'About North-leach the road grows very disagreeable. Nothing appears, but downs on each side; and these often divided by stone walls, the most offensive separation of property.'[46] But this new landscape provided great opportunities for those interested in fox hunting and shooting. For fox hunting the hedges provided the added excitement of many and various jumps for the horses. The layout of the new fields, and the division of estates into newly modelled tenant farms, also provided space for the planting of small woods suitable for pheasant shooting. Landowners enjoyed their ability to roam over the countryside in the hunting and shooting seasons and to share this enjoyment with their wealthier friends and tenants.

Fox coverts were planted to ensure sufficient quantity of foxes in any particular area. In the famous Quorn country of Leicestershire and south Nottinghamshire, for example, 'the making of an adequate number of new covers . . ., appropriately sited and so managed that they would always hold the "raw material" of a hunt', solved the problem of too few foxes.[47] Examples of such coverts include Parson's Thorns and Curate's Gorse, both in the parish of Hickling. Although 'woods and plantations 'afforded 'the warmest shelter' in a cold wind, artificial coverts made 'expressly for the purpose of holding foxes' could be particularly effective. These should be 'not less than five acres or more than twelve' and a 10-acre 'thorn or gorse covert' facing southwest with an edge of 'a double row of Austrian pines, as they grow very quickly and make a splendid break for the wind' would be excellent. Within that there should be a belt of whitethorn and privet with 'the remainder to be divided into four parts . . . two quarters to be planted with blackthorn and the other two sown with gorse.' One of these quarters needed to be cut every three years and allowed to regrow so that the covert did not become 'hollow'.[48] This careful management allowed the covert to provide continual shelter for foxes over many years. And the requirements of the fox for warmth and shelter helped to transform the appearance of formerly open areas by the creation of a scatter of small plantations of gorse, thorn and trees.

The growing enthusiasm for fox hunting in the nineteenth century was more than matched by the increasing intensity of game shooting. The management of woodlands to maximize the number of pheasants available to be shot became of abiding interest to landowners. The preferences of pheasants, which enjoy woodland-edge habitats, increasingly informed the way that woods were managed and plantations laid out. Just as the requirements of the fox led to a pattern of small coverts within a hedged landscape, the needs of the pheasant encouraged the establishment of woods with long woodland edges. The screens and belts around parks were ideal for this purpose, as were clumps and plantations with protuberances.

The demands of the gamekeeper became increasingly rapacious and this sometimes led to conflicts between different departments within the same estate. Tenant farmers were likely to be annoyed by the damage pheasants caused to crops, while foresters often felt that pheasants were more important to the owner than trees. Most importantly, however, the requirements of owners to preserve their game birds led to many often violent conflicts between poachers and gamekeepers. The larger estates had teams of keepers which acted as a local police force to restrict and control all access to estates. The woods where game was reared and fed had always been targets for poachers, and as the numbers of pheasants increased through more intense management, owners became more intolerant of public access to their land.

The types of pheasant shooting changed dramatically over the century. In the early years of the century it was normal to go out, like

Tom and Edmund Bertram in *Mansfield Park*, with a couple of dogs and return happily with a dozen pheasants after a day's shooting. But by the end of the century this was seen as a completely antiquated and outmoded approach:

> in the old days, in big woods . . . the custom was to walk the birds up in a line. But this sort of Early Victorian, pluffing in the tail, jungle-hunting process belonged to a different age, and . . . in a flint and matchlock, or muzzle-loading and percussion-cap age, was governed by different circumstances . . . Under present circumstances such a way of shooting can only be looked upon as a desecration of civilisation, and characterized as a German student's duel, unspeakable and past the pale.[49]

Changes in the design of the shotgun, including breech loading, hammerless guns and cartridge ejectors, increased the accuracy and speed of shooting. The spread of the railway, which made even distant estates accessible, allowed the wealthy to travel to and from shooting parties on a regular basis. The social cachet of the sport was sealed in Britain by the enthusiastic support of the Prince of Wales.

Under the *battue* system, which was influenced by Continental styles of shooting but reached its apogee in Britain at the end of the nineteenth century, rather than walking after pheasants, the birds were driven by teams of beaters over the shooters. Various points around the estate were decided upon as the best places for the shooters to stand, and the woods were designed and managed to maximize the number of birds that could be shot at these sites. This change in style was commented on by Turgenev when he stayed at his friend W. H. Hall's estate, The Cottage at Six Mile Bottom near Cambridge, in October 1878. He had visited the estate to 'compare his experiences of Russian and English sport' and told Tolstoy a month later that he had 'clobbered a fair number of pheasants and partridges, and so on' but complained 'of the monotonous nature of dogless English sport'. On an earlier visit to London in 1857, he visited Joseph Lang's gunsmiths shop in Cockspur Street to order new breech-loading shotguns, and also obtained a pair of lemon-and-white pointer bitches to be sent to Tolstoy.[50]

Sometimes the best design for game was not appreciated in landscape terms and when Hall had inherited the estate he had 'found it laid out with symmetrical coverts and belts designed to create the best conditions for sportsmen . . . he loved the English countryside too much to tolerate this mathematical arrangement. He consequently reshaped the gardens, groves and thickets to be more attractive to his guests.'[51] More usually woods designed for game preservation could also be attractive additions to the landscape. The woods and plantations of the Holkham estate in Norfolk (illus. 88) had

> immense advantages as a pheasant preserve. This is largely due to the arrangement of the woods by the original designers of this demesne. They had a 'clean sheet' to work upon . . . Its fame rests on its good light soil . . . and the massing of the woods round the park, the interior of which . . . forms a great central attraction to game . . . Round this runs a great wood, protected on the outer side by a wall, but having on the inner side this picturesque feature – that in parts the park runs right in among the tree trunks, and one sees hundreds of pheasants feeding in the semi-open space.[52]

The surrounding wall not only kept the pheasants in, but kept unwanted visitors out. The shooting plan has affinities with a battle plan and shows the complexity of arranging several days' shooting and the advantages of clumps, belts, walls and nets to allow pheasants to be held on an estate and presented efficiently, before or after lunch, to the landowner and his friends.

On many estates shooting became the dominant pursuit, even more so with the great agricultural depression that developed from the 1870s onwards, largely as a result of the rapid rise of cheap imports of food from the Empire. For some estates shooting and hunting became their sole *raison d'être*. John Simpson argued in 1907 that 'It is better that every wood and copse on the estate should be a covert, and detached coverts should be as numerous and as widely distributed as extent and circumstances will permit, and should extend to the limits of the estate.'[53] One problem, however, was what to do with large woods from which it was difficult to extract pheasants. One way to deal with them was to cut wide rides through them in which the guns could stand. At Stanage Park in Radnorshire this was done by Lord Powis and 'unwieldy tracts of woodland' which 'used to be shot ineffectually on the sheer haphazard system' now formed 'a most successful and scientific shoot, in which

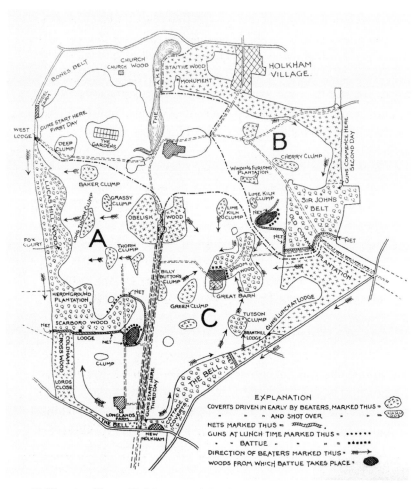

88 'Shooting Plan at Holkham, Norfolk', from Horace G. Hutchinson's
Shooting (1903).

not only do the birds give the best of shots, but nearly all are shown'.
Another way to increase the power of woods to hold pheasants was to
'make a few open spaces and plant the common rhododendron . . . the
rhododendron is about the only positively assured pheasant covert', especially in areas where rabbits would eat everything else. This is one of the
main reasons why rhododendrons, now often seen as a notorious pest
species, were so frequently planted. Another way of improving old
large woods was to cut 'your coppice in strips, so that you have your big
wood composed of coppice of various ages'. But the 'halcyon days when
merchants competed to buy your timber and purchase your coppice at
£9 the acre (double the price of what it is now) – these days are fled'.[54]

Large-scale shooting was a most luxurious and expensive sport and one which could only be provided by the very wealthiest landowners. Towards the end of the nineteenth century Britain's industrial power and Empire, although increasingly threatened, provided enormous wealth for a landowning class that enjoyed conspicuous consumption. Landscapes of covert, clump and plantation were created and managed to provide sport for aristocrats and monarchs. The spectacle of shooting included not only the sport itself but the associated luncheons, dinners and parties. The keepers in special tweeds and the beaters, as in the illustration here (illus. 90), crossing the bridge at Nuneham near Oxford so that they could drive the pheasant and wild duck out from the 'curious horseshoe-shaped islet' covert over the River Thames, became part of an almost theatrical display of power and privilege.[55] Several European monarchs were keen shots, in addition to being targets themselves, including King Carlos of Portugal (assassinated in 1908), who regularly shot at both Windsor and Elvedon; the Maharajah Duleep Singh, with his famous estate in Suffolk; Alfonso XII of Spain; and Kaiser Wilhelm II. When Archduke Franz Ferdinand was at a shooting party at the Duke of Portland's Welbeck estate in 1912 a loader slipped in the snow and accidentally fired two barrels at him but missed by a few feet.[56] One of the effects of the war which followed his assassination soon afterwards was the collapse of the social system that supported such luxurious nineteenth-century woodland sporting landscapes.

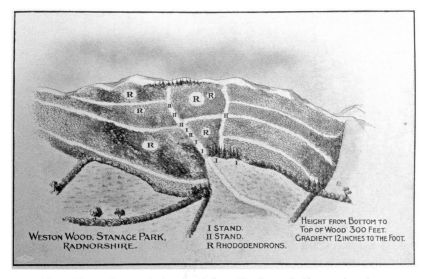

89 'Game Plan at Stanage Radnorshire', from Hutchinson's *Shooting* (1903).

90 'Beaters crossing the bridge' at Nuneham, Oxford, from Hutchinson's *Shooting* (1903).

There were many reasons for the eagerness which many landed estate owners showed towards their woods and plantations in the nineteenth century. For much of the century the woods remained important for coppice and timber, but the commercial production of timber was often not the main reason for establishing new plantations. For many owners there was an enthusiasm for trees that was captured by John Ruskin. He celebrated trees for their 'unerring uprightness', like temple pillars, and 'mighty resistances of rigid arm and limb to the storms of ages'. Trees clothed 'with variegated, everlasting films' the summits of 'trackless mountains' and ministered 'at cottage doors to every gentlest passion and simplest joy of humanity'. Ruskin's admiration for trees went so far that he felt they deserved 'boundless affection and admiration from us', serving as 'a nearly perfect test of our being in right temper of mind and way of life'. No one, he said, 'can be far wrong in either who loves the trees enough and every one is assuredly wrong in both, who does not love them'.[57] And it was to Ruskin that the English Arboricultural Society, founded in 1881 and later to become the Royal Forestry Society, turned for ideas for a suitable motto in 1887. His two suggestions from the Psalms: '*Et folium ejus non defluet*' (His leaf also shall not wither) or '*Saturabuntur ligna campi et cedri Libani quas plantavit*' (The trees of the Lord are satisfied – the cedars of Lebanon which He hath planted) were not taken up, but Ruskin felt that the work of the Society was the 'usefullest of all material work that can be'.[58]

While not everyone would have agreed with Ruskin that affection for trees could be treated as a measure of morality, there was a dynamic zest for planting trees in gardens, parks and plantations. The new plantings, whether individual trees in gardens, parks and hedgerows; fox coverts; small game and landscape plantations; or extensive larch and pine plantations, when combined with existing old woods and coppices, produced a diverse new landscape of trees, woods and plantations. Fragments of ancient woodland jostled with modern mixed plantations and were often amalgamated with them. But within this diversity there were commonalities of estate landscapes. The home woods and parkland near the mansion house would frequently have specimen trees such as cedars of Lebanon and, later in the century, Wellingtonias. The outer shooting coverts spread across tenanted farms had characteristic exotic cover species, such as the Canadian snowberry (*Symphoricarpos albus*), Oregon grape (*Mahonia aquifolium*) and *Rhododendron ponticum*. The larger plantations were most frequently of larch and Scots pine and often mixed with broadleaved trees.

Estates became increasingly private: unwanted visitors were excluded by gamekeepers and, as the screens of trees matured, mansions that had formerly stood out and dominated local landscapes became hidden from view and naturalized. The great wealth of many landowners, derived from urban property, trade and industry, allowed them to treat their estates as sites of conspicuous consumption rather than production. They were usually less concerned with the potential long-term profit from their woods and plantations than with the pleasures that could be gained from their beauty and their crucial importance for delivering exciting fox hunting and scientific pheasant shooting. But the extensive experience of forestry gained by tree nurseries, land agents and owners did provide empirical evidence for the success or otherwise of a wide variety of species under different circumstances, plus a stock of timber that was to be of enormous importance when its importation became almost impossible during the First World War.

EIGHT
Scientific Forestry

I N the seventeenth and early eighteenth centuries there was increasing concern in Europe about how to maintain the supply of wood and timber for burgeoning industries. In Germany the sustainable use of forests was encouraged by the publication in 1713 of Hannss Carl von Carlowitz's *Sylvicultura Oeconomica*, which is frequently identified as the first book published on the economics of forestry. Carlowitz was a mining administrator concerned with ensuring the supply of timber for the mining industry of Freiburg near Dresden. His book was influential in popularizing the idea that carefully regulated management of woodland on a rotation could produce a known amount of wood and timber products into the future. The system depended on diligently mapping woodland and controlling when different sections could be felled, and was dependent on the ownership of a significant area of land. Taking the production of firewood from a mixed coppice of ash and maple as an example, this could be cut on a twenty-year rotation. If you wanted to ensure a constant supply of firewood, you had to divide your existing woodland into twenty compartments and one-twentieth of the whole woodland could be cropped annually. At the end of twenty years all the woodland would have been cut once, and the rotation could be started again. Clearly this only worked well if the total area of woodland was sufficient to fulfil your firewood requirements: if the demand for wood was increasing you would need to increase the area of woodland you controlled proportionately.[1]

Early 'sustainable forestry' in Europe

One way to increase the productivity and profitability of areas of woodland was to control traditional uses such as grazing, which might adversely affect the regrowth of coppice. In much of Europe forests had long been used to produce firewood and timber, but these uses were somewhat secondary to the agricultural uses of the forest: namely the grazing of wood pastures and the browsing of trees by stock, which allowed large flocks and herds of animals to be maintained. These were crucial for the agricultural economy in producing meat, milk, leather, wool and fertilizers. As in many parts of the world, the form of woodland management was strongly influenced by the balance between communal rights over land and more centralized control by both private and state landowners. Communal rights to grazing and pasture gradually began to be seen by many landowners as a significant deterrent to the introduction of more productive and sustainable forms of woodland management.

It is only with the removal of conflicts with game and agricultural uses in the eighteenth century that the new forestry techniques associated with forestry systems such as *Schlagwaldwirtschaft* could be established. This system of growing and felling trees using some sort of rotation, section by section, had as characteristic features 'sustainability, a scientific approach, central management and the primary orientation towards wood production'. Research in the forests of Hunsrück and Eifel demonstrates that some of the key characteristics of scientific forestry were developed in the eighteenth and early nineteenth century. When individual landowners had gained full control over grazing, other agricultural practices and hunting, they had the confidence to develop new forms of woodland management. The obverse of new forms of woodland management was the removal and extinction of various common rights. From 'this time on "sustainability" was something like the pivotal point of forestry' and 'the economist Pfeiffer in 1781 saw a way to realize 'the perpetual maximum utilization of the forests.' But how was this new programme of controlled forestry to be enforced? The answer was to establish detailed and clear forest laws and an uncorrupted forest administration of trained forest officers and staff to ensure they were followed. This authoritarian approach was followed and existing laws dating back to the medieval period were codified precisely. A new forest administration with foresters trained in the science of forestry was established. A key point was the need for this administration to separate itself from the traditional worlds of hunting and agricultural exploitation of the forests. The large herds of deer favoured by princely families and their households had to be reduced as these hindered regrowth of forests, and grazing by sheep and cattle began to be suppressed.[2]

But there is also clear evidence that tracts of woodland were managed in a consciously sustainable manner much earlier than the eighteenth century. Richard Keyser's analysis of monastic records in the Champagne region of northern France provides an excellent insight into early developments in sustainable woodland management in Europe. He argues that 'high medieval demographic and economic growth in Champagne and other parts of northern France encouraged a switch in the primary focus of woodland management from grazing, hunting and other relatively extensive methods of gathering naturally occurring products to intensive small wood production'. This was stimulated by rising urban demand traces of which can be seen in surviving monastic records which have an increasing number of 'commercial contracts based on coppicing'. Before the twelfth century the evidence suggests that most woodlands managed by great landlords were *silva glandaria*, which produced timber, acorns and beech mast. Tenants paid these landlords 'dues for wood gathering and for pasturing pigs'; they also provided labour services by 'cutting and hauling both small wood (*lignum*) and building lumber (*materiamen*)'. Coppicing clearly took place, but there are only a few explicit mentions in the documents, such as the reference to an estate owned by the abbey of Saint-Remi of Reims near Châlons, where 'a tenant family held, along with six units of arable, three of *silva minuta*, which produced small wood for fencing.'[3]

Monastic records indicate that there was considerable regulation of the management of trees and woods after around 1170. As the populations of towns such as Brie in southern Champagne reached 10,000 and that of Troyes around 20,000, the demand for firewood and construction materials encouraged 'intensive woodland management in the thirteenth century'. Greater efforts were made to enforce long-established rules to keep pigs from damaging coppice regrowth, and to mark ownership boundaries more precisely. Regulations controlling the extraction of wood began to appear near cities: in 1197 an early reference to the commercial supply of wood is shown by the villagers of Fays being allowed to sell small wood in Troyes, 20 kilometres (12½ miles) away. The growth of this trade is indicated by actions of the Benedictine monks of Montier-la-Celle, just outside Troyes, who tried to control the use of their woods at Jeugny, 18 kilometres (11¼ miles) to the south. An accord of 1220 with the nuns of Notre-Dame-aux-Nonnains of Troyes limited the amount of firewood they could collect from the Jeugny woods: 'The nuns will have a single wagonload [*biga*] of wood pulled by two horses per day through the year, . . . for all types of wood except standing oak and beech, . . . they can have oak and beech that is lying on the ground, as long as it is neither fit for lumber nor part of an ongoing sale.'[4]

Specific contracts for wood cutting started to become common in the early thirteenth century. One of the first to demonstrate the division of woodland into sections is a sale by the Benedictines of Molesme, approved by Countess Blanche of Champagne in 1219, of 1,000 arpents (about 2,000 acres) of woodland near Jeugny, south of Troyes. The large-scale nature of this contract is shown by the fact that the two purchasers, Girard Judas and Guillaume de Vaudes, gained a ten-year lease and had to cut 100 arpents of woodland a year. Other contracts approved by Blanche include one of 1217 where 'the countess sold cutting rights' in two small forests or '*forestellas*' over a six-year period 'on condition that the merchants "cut each tree only once so that it grows back quickly"'. In another contract it was specified that 400 arpents would be cut over a ten-year period in the forest of Gault, and adjoining parcels had to be felled in sequence. In addition the restriction on grazing became more closely formalized and regulated. Restrictions on the grazing of freshly coppiced areas for a period of between four and six years after cutting were commonplace by the mid-thirteenth century. In 1271 the Grand Jours of Troyes, the high court of Champagne, 'upheld against the community of Chaource a customary exclusion of pasturage (*vaine-paturage*) for five years after cutting' which allowed the woodland 'to defend itself'. This exclusion period was extended to six or seven years for woodland on poor soils where the coppice regrowth was likely to be less rapid.[5]

In 1284 Philip IV acquired the province of Champagne by his marriage to Joan of Navarre and Champagne. He kept on the existing forest manager, Pierre de Chaource, whose *Book of Sales of the Woods of Champagne* describes the extent and importance of commercial coppicing in the thirteenth century. The book includes details of over 200 wood sales made between 1280 and 1300, mostly at Villemaur in Othe forest, about 30 kilometres (18½ miles) southwest of Troyes. The book names the seller, the area of woodland (usually around 40 arpents) and the price fetched (around £6 per arpent). The income from these woods was enormously important, bringing in about £1,300 per year, which was 'an amount similar to that of the city of Troyes's annual tax'. The details of the contracts show that purchasers normally had six years to coppice their plots and could pay for this in annual instalments. Payments were usually made on 30 November, St Andrew's day, which marked the start of the main coppicing season. The cutters were a mixture of local people and merchants from Troyes who might make several purchases. Local villagers had a keen interest in checking that the contracts were scrupulously followed, and a range of officials was employed that included forest managers (*gruyers*) such as Pierre de Chaource, guards, sergeants and, perhaps most important, surveyors (*arpenteurs*) who marked out the areas to be sold and coppiced (illus. 76).[6]

Although coppicing is an ancient practice of great significance and importance in many parts of the world, it is in the later Middle Ages that it can be documented as being of commercial importance. It was in the thirteenth century that it began 'to dominate sylviculture in parts of northern France as a market-orientated system. At once commercialized and sustainable over centuries, this system would persist into the nineteenth century.' Many French forest historians have remained 'resolutely modern and statist' and focussed on how later national governments 'gradually imposed order and a focus on timber production on earlier, insufficiently regulated practices'. In so doing they have missed the documented scale of the extent and commercial importance of coppicing in this earlier period, when the 'coppicing cycle was clearly the primary regulator' of woodland management, with the production of timber from standard trees (*bailivaux*) being of secondary importance. The sustainable nature of 'intensive sylviculture based on coppicing' was recognized: a royal decree of 1346 stated that masters of forests where sales were to take place should visit them and 'inform themselves about all those forests and woods' so that 'the said forests and woods can be perpetually sustained in good condition'.[7]

While hunting is often identified as a way of displaying power and wealth, its role as a mode of surveillance is probably underestimated. Hunting over their lands allowed the great landlords, princes and bishops of the medieval period and later to ascertain the condition of their woods and the quality of the work of their agents, foresters and woodwards. Christoph, Duke of Württemberg (*fl.* 1550–68), for example, was an enthusiastic regulator and administrator who was keen to extend control over his woods. Attempts 'to improve the quality of the woodlands between Stuttgart and Boblingen' followed 'critical comments made by Duke Christoph during a hunting trip in the area in 1564'. Hunting also gave additional influence and authority to forest officials and administrations because they worked closely with the centres of power. Joachim Radkau has noted that some 'historians think it is largely owing to princely big-game hunting during the early modern era that many woodlands were saved from deforestation. At least it was the deer-hunt which gave power and prestige to several forest administrations' in the seventeenth and eighteenth centuries.[8]

The idea of sustainable woodland management systems has a long history, stretching back to the medieval period and often involving woodcutters and merchants, landowners and forest officials in complicated legal agreements. Indeed, it is largely through these surviving legal documents that the details of woodland management history can be envisioned. Sustainable cutting of coppice undoubtedly has existed for much longer than the surviving documents allow us to determine. However, there is little doubt that the rise of what can be described as modern scientific sustainable forestry took place largely in German states. Radkau points out that *Nachhaltigkeit*, or sustainability, has in recent years 'become a magic word in German forestry' and that the 'establishment of the principle of *Nachhaltigkeit* is usually claimed as the great historical achievement of Germany which spread from Germany all over the world'. He argues that the word is often understood in a 'merely quantitative, mathematical manner', showing that a forest policy can guarantee the regeneration of a certain amount of wood in a certain time. But it is the apparent certainty and security of future wood supplies provided by the careful calculations and accounts which made this new scientific forestry so attractive and beguiling.[9]

Scientific forestry in Germany, India and America

A key difference between earlier concepts of sustainability, which were usually concerned with the careful management of coppice woodland under rotation, and the sustainability of the new scientific forestry was that the latter is associated strongly with the afforestation of former cleared land. From the sixteenth century onwards there had been concerns over a timber famine in Germany, as in many European countries, and this became a strong argument for the establishment of new areas of woodland. It was this policy of reafforestation and the establishment and management of high forest (*Hochwald*) that German foresters became famous for in the early nineteenth century. One of the most influential forest scientists was Heinrich Cotta (1763–1844), who founded the forest school at Tharandt, near Dresden, that became the Royal Saxon Forest Academy and was enormously influential in training foresters; and another was Georg Ludwig Hartig (1764–1837), who was chief inspector of forests at Stuttgart and later Berlin. There was a very strong demand for timber of high quality grown over a long rotation, especially through the Dutch timber trade. But there was another reason for the rapid development of the new scientific forestry profession: 'only a policy of high forest with long cutting cycles was able to justify an independent and well-established forest administration and to defend it against a rising tide of liberalism, which originally was opposed to governmental forest administration.' Whatever the motives, the rise of scientific forestry transformed the public image of foresters, who by the mid-nineteenth century had become 'one of the highest-esteemed German professions and were regarded as defenders of nature, advocates of the common wealth and the interest of future generations'. The German writer Friedrich Schiller, who had at first thought of foresters merely as hunters, 'developed a high respect for the profession when he heard that Hartig made forest plans for more than 120 years ahead'.[10]

The influence of German scientific forestry began to be felt in the English-speaking world from the mid-nineteenth century onwards. The ideas did not travel directly across the English Channel to Britain, but went along an indirect route through British India, where the need for specialized forestry personnel was recognized early in the century. While the 'first attempts at forestry conservancy' began in Burma as early as 1826, and Mr Conolly the Collector of Malabar 'commenced planting teak on a large scale at Niambur' in the 1840s, it is to a later generation of foresters that the instigation of scientific forestry can be attributed. The Earl of Dalhousie, Governor General of India from 1848 to 1856, was a keen administrator and issued a Charter of Indian Forestry in 1855, which established that land not privately owned was state land and 'with the establishment of forest areas as absolute state property, the Charter required proper management of the forest areas and this meant scientific forestry.'[11]

Two of the most important figures in the promotion of scientific forestry were the German botanist Dietrich Brandis (1824–1907) and the German forester William Schlich (1840–1925). Both men made their careers in India and Britain and were eventually knighted for their services to forestry. Brandis became interested in botany as a child in Athens, where his father worked for King Otho, and gained a PhD at

Bonn. The decisive link to Indian forestry came with his marriage to Rachel Havelock, whose brother-in-law the Indian army officer General Henry Havelock recommended Brandis to Lord Dalhousie. Brandis was appointed in 1856 to put a stop to illegal fellings of valuable teak trees in Pegu and his success in this task and his administrative skills meant that he was appointed inspector-general of all Indian forests by 1864. He visited many Indian forests during the 1860s and his reports showed 'a keen interest in evaluating community forest management. In the debate among officials prior to the second Forest Act of 1878 he steered a middle course between advocates of total state control of forests and votaries of village control.' William Schlich was trained as a forester and after graduating worked for the Hesse state forestry service but he lost his job following the Austro–Prussian War of 1866 and moved to India in 1867. He first worked in Burma and later in the Punjab and succeeded Dietrich Brandis as inspector-general in 1881. While in India he established the 'imperial working plans branch, which ensured the preparation of forest working plans on approved lines and their scrutiny by a central authority.'[12]

While both Brandis and Schlich made their careers and gained their practical experience of implementing forest plans in India, it is through their work in education and writing that they had their greatest influence across the world. In order to gain well-trained forestry officers in India, from 1866 onwards 'a number of selected Englishmen were sent annually

91 The tropical forestry expert Dietrich Brandis with students at the University of Giessen, Germany, in 1889.

for a term of two years and eight months to the Continent to study Forestry, half of them going to France, and the rest to Germany.' Brandis later urged that some form of British school of forestry education should be established and in 1879 the Forestry School of Dehra Dun was opened.[13] This was later reorganized by Schlich, who then left India to establish a forestry department at the Royal Indian Engineering College at Coopers Hill, Englefield Green, Surrey. His former colleague Brandis, back in Germany, 'agreed to supervise the practical continental training' of students who were mainly British, or from the Empire, but also included several Americans. This college closed in 1905 and the training of foresters was taken over by the University of Oxford. Schlich was responsible in both institutions for 'the training of no fewer than 272 out of a total of 283 officers who joined the Indian forest service during that period'.[14]

William Schlich's massive *A Manual of Forestry* was published in five volumes from 1889 to 1895. Its gestation and form demonstrate how the ideas of scientific forestry which originated in Germany flourished in India and then became distributed around the world. The production of a series of handbooks on forestry was a key part of Schlich's role at Coopers Hill. Although he did draw on some books by British authors, such as Brown's *The Forester*, the main intellectual thrust is from German authorities, including Schwappach and Baur on forest mensuration, G. Heyer on forest valuation and Friedrich Judeich's *Die Foresteinrichtung* on forest working plans; the last's 'method of regulating the yield of forest' Schlich thought the best, though sometimes 'too rigid'.[15] In the third volume of the *Manual* he made use of forest working plans provided by German colleagues for Krumbach Communal Forest and the Herrenwies Range, the latter prepared for him when he toured the forest with Coopers Hill students in 1893. The final two volumes are not by Schlich but by his colleague at Coopers Hill W. R. Fisher: *Forest Protection* was adapted from Richard Hess's *Forstschutz*, while the fifth volume on *Forest Utilisation* is a translation of *Die Forstbenutzung* by Karl Gayer, first published in 1863.[16]

Schlich's *Manual* was only the largest of a number of important books published in the late nineteenth century popularizing German forestry. John Nisbet of the Indian Forest Service published *Studies in Forestry* in 1894, noting that they were based on his 'Essays on Sylvicultural Subjects', which were 'written by me in Bavaria during 1892' and published by the Government of India in 1893 'for distribution among their Forest

Officers'. Nisbet noted that 'it will be apparent throughout every chapter' that 'my convictions regarding economic Forestry . . . have been formed in a Teutonic school.' He implicitly identifies a tension in using German models to further British forestry and argues somewhat defensively that 'in acquiring information with regard to the growth of Forest Trees . . . it ought to be a matter of perfect indifference from what well this may be drawn.' He therefore 'had no hesitation in boldly acknowledging the German sources from which many of the lessons I am trying to teach have been learned'.[17]

The methods and ideas of scientific, sustainable forestry spread rapidly through the British Empire, especially in Australia, New Zealand, South Africa and Canada, and this was furthered in the twentieth century by the establishment of new forestry schools and colleges and the holding of regular British Empire Forestry Conferences from 1920 through to 1947.[18] But the links with the USA were very strong and American foresters were also strongly influenced by the work of Brandis and Schlich. One of the leading proponents of working plans for American forests was Charles Sprague Sargent (1841–1927). Sargent was a botanist who directed the Arnold Arboretum at Harvard for 54 years and published many works, including *The Silva of North America* (1891–1902) in fourteen volumes and the *Forest Flora of Japan* (1894). He was also chairman of the National Forestry Commission, which surveyed the nation's timber reserves. In his popular magazine *Garden and Forest*, published 1888–94, he argued that 'India has given to the world the most conscious example of a national forest policy adopted over a vast area' and recommended the adoption of 'empire forestry-style working plans adapted to the American market'.[19] Sargent was delighted when Dietrich Brandis was asked by the National Academy of Sciences to produce a 'plan of action for the protection of American forests'. Sargent thought that 'it ought not to be impracticable to frame a system of forest management' for America 'which would contain all the essential features of the plan which has proved such a conspicuous success in India'.[20]

Gifford Pinchot (1865–1946) was the most important figure in the development of forestry in America. He was appointed first chief of the Forest Service (1905–10) and before that had been chief of the Division of Forestry from 1898 to 1905. There was no university training for foresters in the States and after Yale he 'went to Europe to study under French and German *trained* foresters, who in turn had served much of their career in British India'. When in England Pinchot was advised by William Schlich to 'strike for the creation of National Forests' back home and, with 'a copy of Schlich's first volume of *Manual of Forestry* under his arm', travelled to Nancy, where many Indian foresters had been trained. He then studied in Germany with Brandis, who was a strong and continuing influence. Pinchot was impressed by the '"multi-use" model of forest management' that Brandis had devised for British India, which 'reconciled the needs of peasants, businessmen and environmentally prone administrators'. When Pinchot returned home he worked on the Biltmore Forest Estate near Asheville, North Carolina, owned by the enormously wealthy George W. Vanderbilt, for three years and later worked with the National Forest Commission of the National Academy of Sciences and travelled extensively to identify possible forest reserves. His appointment as chief forester allowed him to increase the area of national forests from 56 million acres in 1905 to 172 million acres in 1910 and in this he was helped by the support of his friend President Theodore Roosevelt. Pinchot noted that when he 'came home not a single acre of Government, state, or private timberland was under systematic forest management anywhere on the most richly timbered of all continents'. He argued that 'the common word for our forests was "inexhaustible"' and that lumbermen 'regarded forest devastation as normal and second growth as a delusion of fools . . . And as for sustained yield, no such idea had ever entered their heads . . . What talk there was about forest protection was no more to the average American that the buzzing of a mosquito, and just about as irritating.'[21]

One of the most colourful German foresters working in America in the early twentieth century was Carl Alwin Schenck (1865–1955), who gained his PhD in forestry at Giessen. He knew Brandis in Germany and served as William Schlich's assistant on European tours by English and Indian forest students from 1892 to 1894. In 1895 he moved to become forester at Vanderbilt's Biltmore estate after he and Pinchot remembered that it was Frederick Law Olmsted, who had designed the landscapes at Biltmore, who had persuaded Vanderbilt to take an interest in forestry and invite Pinchot and Schenck to develop forestry there. He worked as forester to Vanderbilt until 1909, introduced forest-management plans and founded the Biltmore Forest School, which laid claim to be the first forest school in the States, in 1898. Before it closed in 1913 around 400 students graduated in forestry, many going on to establish modern forestry practice in America. Schenck served

92 Forestry educators William Schlich (front, centre) and Carl Schenck (front, right) with students in Saxony, 1892.

with the German army on the Russian front during the First World War and lived in Germany afterwards, leading many forestry tours in Germany and Switzerland in the 1920s and '30s.[22] Once forestry training started to become established in America, young American foresters went to Germany to learn European forest-management methods. An important example is provided by Arthur Recknagel (1883–1962), who had worked in the forestry service from 1906 to 1912, when he went to Germany to study forest management for a year. He was Professor of Forest Management and Utilization at Cornell from 1913 to 1943; Cornell had been established in 1898, the same year as Biltmore Forest School. He wrote in the preface to his book *The Theory and Practice of Working Plans* (1913, second edition 1917) that he presented 'the best European efforts' on forest organization and planning 'adapted to the present needs of American forestry'. Prominence is given to examples of working plans from Prussia, Bavaria, Saxony, Württemberg, Baden and Alsace-Lorraine, with some examples also from Austria and France.[23]

The new forestry in Britain

The practices of forestry and woodland management were transformed in late nineteenth- and early twentieth-century Britain. On the one hand, there was a long, quiet decline in traditional woodland-management techniques, such as coppicing, which went largely unnoticed by contemporaries. On the other, there was an orchestrated campaign to form what was termed a 'new forestry' based on scientific principles developed on the Continent. While British forestry policy in India was central to the spread of scientific forestry in the second half of the nineteenth century, there was little evidence of the successful introduction of these new approaches in Britain itself, where the interests of landowners tended more to hunting, shooting and the appearance of the landscape than the efficient production of timber. Many in Britain argued that one reason for the lack of a sustained and economic forestry was the lack of state forestry. Some foresters looked with envy at European state and communal forests, where long-term experiments and schemes for improved forestry could be instituted.[24] Schlich had great influence on the orientation and development of forestry education in Britain in the Edwardian period, especially through his *Manual of Forestry*. Lectureships in forestry were established at the University College of North Wales at Bangor and at the Armstrong College at Newcastle upon Tyne in 1903. A School of Forestry for Woodmen at Parkend in the Forest of Dean was opened in 1905, and the following year Schlich transferred the forestry section of the Royal Engineering College to the University of Oxford. A new generation of trained foresters was produced whose knowledge had been imported from Germany and France and who were essentially concerned with plantations rather than with coppice with standards.

This circulation of ideas, formal and informal, and its implications for British forestry was recognized by A. C. Forbes, lecturer in forestry at Newcastle in 1910, who pointed out that 'since about 1860', with the establishment of the Indian Forest service, 'a small stream of continental trained youths has been going out to India, an equally small stream of retired Indian foresters . . . has been returning from it.' He argued that 'Whatever the exact practical results of this inter-mixture of British and Anglo-Indian ideas may have been, there is little doubt that fresh ideas were instilled into British foresters and proprietors, and a wider knowledge of forestry as an industry instead of a hobby resulted.' Formal education was not, however, the only way in which Continental scientific modes of forestry were popularized in England: 'the constant visits made by the British landowning class to the Continent in search of pleasure, sport or health', although not directly connected with forestry, 'can scarcely have failed to open the eyes of landowners to the possibilities of scientific forestry'.[25] These novel modes of scientific forestry had to compete with the powerful interests of estate owners in shooting and hunting, which had for many years co-existed with traditional coppice woods and woods consisting of coppice with standards.

A good indication of the increasing interest in scientific forestry was the establishment in 1882 and increasing popularity of the English Arboricultural Society (later the Royal Forestry Society). There was considerable interest in making plantations on agricultural land of poor quality, including heaths, moors and sand dunes. The first direct impetus for state forestry in the Edwardian period came from an unexpected source. In 1909 the Royal Commission on Coastal Erosion reported. One of its conclusions was that 'whether in connection with reclaimed lands or otherwise, it is desirable to make an experiment in afforestation as a means of increasing employment.' In the same year, the Chancellor of the Exchequer, David Lloyd George, announced in his budget speech that money was to be made available for schools of forestry, the acquisition of land for planting and the creation of experimental forests. An advisory committee on forestry was appointed in 1912 by the Board of Agriculture and its report, published in the same year, recommended that a forest survey should be carried out and that 5,000 acres of land should be acquired for such an experimental forest.[26]

Many British foresters, however, remained uncertain of some of the benefits of scientific forestry and a debate in the *Quarterly Journal of Forestry* of 1914 highlights this conflict. The forester Thomas Bewick, writing on a successful visit of British foresters to Germany in 1913, noted that 'we cannot under present conditions adopt the German system of forest management in extenso, but I am sure that some thing could be done in a modified way.' William Schlich, at the age of 74 the doyen of the British forestry establishment, did not allow Bewick to get away with this rather lukewarm appreciation of the merits of scientific forestry. In the following issue of the journal he argued that

> it has often been said that the continental systems of forest management are not much use in these islands, because here

entirely different conditions have to be dealt with. I do not think any sensible person has ever suggested that we should adopt the continental systems en bloc, but [we should] . . . profit by experience in other countries.

Schlich's main argument was that British foresters must stop their tradition of 'haphazard procedure' and start making use of long-term working plans. From such plans, he noted, 'no deviation' is allowed 'except with the sanction of higher authorities'. Moreover, 'in every case it is strictly laid down what is to be done during the next ten, or in some cases twenty, years.' His conclusion was that the 'preparation of well considered working plans is a crying necessity in all British forests, even if they be of moderate size'.[27]

It is difficult to identify the precise impact of these Continental ideas on English forestry in the early twentieth century. Many of the larger estates employed forestry specialists to assess their woodlands and possibly draw up plans. In practice, however, the foresters and woodmen employed by estates were still untrained in the new methods and ideas of forestry. When Sir William Schlich 'visited private estates in England in order to provide working plans for them, he tried to encourage the idea of systematic forest management which he had been taught in Germany'. Later, however, he regarded this work 'as some of the least successful of his career and he attributed this to the frequent changes in the ownership of private estates and to the absence of any tradition of forestry' in England. He made more direct impact through his teaching at Oxford, and Roy Robinson (1883–1952), an Australian Rhodes Scholar with 'athletic prowess' whom Schlich considered his 'most brilliant student', later became a key figure in the development of British forestry. After gaining his Diploma in Forestry at Oxford he became an inspector in the Board of Agriculture and Fisheries in 1909, with responsibility as a forestry advisor. He carried out 'intensive surveys in Wales and the north of England' which, together with 'several extensive motorcycle tours of Scotland', gave him a 'wide knowledge of the growth of trees in Britain'. This was the first appointment of a trained forester by the Commissioners of His Majesty's Woods, Forests and Land Revenues, and gave him oversight of the management of the remaining Royal Forests, including the New Forest, the Forest of Dean and other smaller but still significant Crown holdings. These estates were 'administered by several firms of land agents while the woods were locally supervised by Crown Foresters', a mode of control later described by the forester George Ryle as 'a combination of entrenched conservatism that was to take a great deal of breaking down'.[28]

Roy Robinson was identified as the right man to make sweeping changes: he was 'severely practical' and 'tried most operations himself and abandoned anything not strictly necessary'. Moreover one of his 'chief attributes was tenacity of purpose' and his 'approach to forestry was mathematical and detailed'. But his scope of action with the Crown Forests was relatively limited. The lack of any significant 'higher authorities' who could lay down rules that private landowners would have to follow was, perhaps, the main reason why modern forestry did not gain a significant foothold in Britain before the First World War. The tremendous social and political changes associated with the onset of that war, however, allowed the idea of a state forest service to become credible, if not welcomed by all. A debate held by the Royal English Arboricultural Society in September 1916 demonstrated the divisions between those members who saw a state forest service as essential to the development of British forestry and those who saw it as an attack on the freedom of private estate owners. By the end of the war, however, the establishment of a state forestry service was generally welcomed as a necessity. The luxurious forestry traditionally practised by the landed estates was recognized as anachronistic, and the need for a strong state forest organization to produce large quantities of timber and a modern forest landscape was seen as essential.[29]

The First World War and British forestry

The First World War shattered the confidence of the landowning classes in Britain. In the Edwardian period private landed estates and the rural way of life associated with them had symbolized the power of private property and aristocratic values. For many people the country-house weekend, shooting parties, hunting and fishing epitomized social success. In practice, however, the power of the landowning classes had started to diminish in the late nineteenth century. The agricultural depression, whose most intense period lasted from the late 1870s to the late 1890s, led to a dramatic fall in estate rentals and, correspondingly, land values. Landed wealth could no longer compete with industrial wealth, and the introduction of estate taxation reduced what wealth remained. Increasingly landowners began to sell off their land. This loss of income was accompanied by a decline in the political power of the aristocracy brought about by major changes to the electoral system. The number of men who could vote increased from around three million men in 1883 to nearly six million in 1888, and in 1918 universal adult suffrage for men was introduced. The final blow for many landed estates was brought about directly by the war as hundreds of heirs to estates were killed. Many landed estates were put on the market, and in 1919 there were some half a million acres of English land up for sale, with twice as much actually being sold.[30]

Landowners, estate staff and farm workers were all directly affected by the horrific levels of death and injury in the war. The destruction of large tracts of countryside along the Western Front in Europe became for many potent symbols of the destruction of life and civilization. Paul Nash's haunting images of shattered trees, such as his *We Are Making a New World* (illus. 78), are 'terrible images' because of 'their combination of detached, almost abstract, appreciation and their truth to appearance'. Nash had been at the Ypres salient in 1917, and after being invalided out returned to the front as an official war artist. The drawings he made in the field of 'shorn trees in ruined and flooded landscapes' struck a chord with the public and made his reputation.[31] The oil painting *We Are Making a New World* is based on his watercolour *Sunrise: Inverness Copse*, a battlefield drawing made at the scene in 1917.[32]

The war not only destroyed trees and woodland in France, but consumed huge quantities of timber for the construction of trenches, walkways and roads. Vast quantities of timber were felled in the immediate countryside, but also shipped and moved by rail across the Continent. Soldiers would not only have been aware of the dramatic destruction of woodland on the battlefields, but also in the woodlands of France, Belgium and Germany. When they returned home on leave, they would also have seen the vast acreages of devastated woodland that were felled for the war effort. Enormous pressures were placed on existing woodlands and plantations and approximately 450,000 acres of woodland across Britain were felled. At the outbreak of war in 1914 there was 'no immediate anxiety' about the supply of timber and 'it was not until 1916 that vigorous and near-panic action had to be taken' to maximize the home production of timber as a replacement for the increasingly perilous supply of timber from overseas. One of the most important requirements was timber to be used as pit props in the coal mines which were so crucial for the British Navy. Eventually the urgency of the timber requirements resulted in the imposition of state control over timber felling and supply.

A Home-grown Timber Committee, chaired by Liberal MP Francis Dyke Acland, was established in November 1915 and this worked until March 1917, when its duties were transferred to the Directorate of Timber Supplies set up by the War Office.[33] This had the enormous job of selecting, felling and controlling the supply of timber from across Britain. Many of the key posts in the department were 'filled by university-trained foresters'. By the end of 1917 the Department 'was already running 182 sawmills' supplemented by an additional 40 mills run by 'a Canadian Forestry Corps, a New England Sawmill Unit and a Newfoundland Forestry Corps'. Much work was also done by the Women's Forestry Corps, and by 1918 around 15,000 staff were employed in timber supply.

The necessary frenzy of timber production brought to everyone's attention the need to consider forestry policy after the end of the war. How should the felled woodlands be replenished? Should vast new areas of land be afforested to ensure future timber supplies in case of national emergency? How could landowners be encouraged to manage their woodlands effectively? The Reconstruction Committee set up a sub-committee on forestry in 1916, also chaired by Francis Dyke Acland, and the secretary to this committee was the dynamic Australian forester Roy Robinson. The remit of the committee was 'to consider and report upon the best means of conserving and developing the woodland and forestry resources of the United Kingdom, having regard to the experience gained during the war'. The membership of this committee included a mix of landowners with forestry interests, government officials and the omnipresent Sir William Schlich. The report made a strong case for the establishment of a national forestry policy. The main arguments put forward were that the 'dependence on imported timber was a grave source of weakness in time of war' and that it was precarious 'even in peace conditions' to rely on securing softwood requirements from overseas', particularly from what were termed 'extra-Imperial countries'. A third argument was that there were large areas of moorland and heathland that were no more than 'waste', and that if these were afforested it would help to stem rural depopulation and increase the general productiveness of the land, since forest establishment and management employed more people than hill sheep farming.

The Acland Committee recommended the establishment of a centralized Forestry Commission, which should be responsible for afforesting and encouraging others to afforest 1,770,000 acres of land over an 80-year period, leading roughly to a doubling of the area of woodland in the UK. It was envisaged that 200,000 acres would be afforested within ten years, '150,000 acres by direct State action' and the remainder largely by private enterprise. The report did not flinch from pondering the difficult problem of the profitability of afforestation. It argued that 'from a national point of view' direct profitability was not the most important aspect and 'although much discussed in this country' it had never 'been so regarded in the countries where silviculture has been longest practised and is most valued'. It considered that 'direct gain or loss is relatively a small matter compared with the new values created', which were expressed 'partly in population and partly in terms of wealth'. The committee concluded triumphantly that 'It is on such values that the strength of nations depends.' But the Treasury representative on the committee, Mr L. C. Bromley, had some strong reservations. He agreed that the proposed commission 'would provide the best means of securing the conservation and development of the woodland and forestry resources' of the country, particularly 'as a safeguard in the event of a possible future war on a similar scale to the present'. But he noted that the 'expenditure proposed is very considerable and that the financial results are likely to be unfavourable to the Exchequer'.[34]

After the publication of the report, things moved very fast, with an Interim Forest Authority established under Acland in 1918. Following the Forestry Act of 1919, the Forestry Commission was established on 1 September 1919. Although severely threatened by the Geddes Axe of 1922, which 'argued that the afforestation policy adopted by Parliament should be completely scrapped', the Commission survived, and in its first ten years very nearly achieved the target of 200,000 acres of woodland establishment. Most of the land purchased or leased by the Forestry Commission was marginal land of relatively little importance for agriculture, such as low-grade grazing land. Land not under economic production was considered 'waste'. George Ryle noted that land chosen for afforestation included 'heathlands and other really unused wastelands such as those in Hampshire, Dorset and more particularly the Brecklands of East Anglia. There were sand-dunes on the coasts at Culbin, Pembrey and Newborough. There were the bracken and heath wastes of the Sherwood Forest.' In addition there were the huge areas of recently felled and unmanaged 'derelict or nearly derelict woodlands'.[35]

About 162,000 ha (400,300 acres) of new forests had been planted and purchased by the Forestry Commission in Britain in the interwar period and over 4,000 people were employed by the Commission. The new plantations were celebrated by many as the efficient conversion of wasteland into modern forests which would provide rural employment and a strategic reserve of timber. The enthusiasm and experience of the nineteenth-century tree enthusiasts and arboretums in testing introduced trees became of vital importance. Sitka spruce was seen, for example, as an ideal tree for upland areas, being 'a substitute for Norway spruce at high altitudes or wherever volume production is more important than quality of timber'. It was favoured as it produced 'timber much more rapidly than Norway spruce and under a wider range of conditions'. Experiments were made to establish the best species to use on a variety of soils and sites. In southern and eastern England the chalk downs and wolds were identified as 'well suited to afforestation' and from 1927 the Commission carried out experiments to discover 'the most suitable kinds of trees to plant' on these chalky soils. A variety of pioneer trees such as Scots and Austrian pine, European larch and Italian alder (*Alnus incana*) were tested, as were different mixes of beech with conifer nurses such as *Thuya plicata* and Lawson cypress, with the general aim of achieving a final crop of beech. These experiments drew on historical knowledge of similar plantings made by private estates.[36]

The effect of the war on British forestry was transformative. The irreversible decline in the power of the large landowners was hastened; many large landed estates were sold and traditional forms of woodland management began to disintegrate. The state Forestry Commission acquired huge tracts of land, planted them uniformly with conifers and introduced grant aid for the management of private woodlands. The concept of scientific forestry finally gained a foothold on British soil. The thousands of acres of coniferous trees that were planted not only formed the kernel of a strategic reserve of timber but challenged traditional landscape values. The geometric form of the new, extensive plantations marched starkly across the semi-natural moors and heaths. The rise of modern forestry obscured the decline and finally the collapse of traditional woodland practices, which would not be revived until interest in conservation management gathered momentum at the end of the twentieth century.

NINE
Recreation and Conservation

I N Alfred Hitchcock's film *Vertigo* (1958) the character played by Kim Novak visits a grove of coast redwoods (*Sequoia sempervirens*). Novak plays the role of Madeleine, a character apparently besotted with Carlotta, a woman who is long dead. She is struck in the film by the cross-section of a redwood which had been felled in the 1930s and displayed marked up with key dates in history. The tree is dated back to AD 909 and the associated dates appear to the modern eye remarkably Anglocentric: 1066, Battle of Hastings; 1215, Magna Carta Signed; 1776, Declaration of Independence (illus. 94). But the character, who is trying to impress James Stewart's 'Scottie' of the veracity of her infatuation, appears more interested in the way she can place Carlotta's life in a specific historical context. In 2003 Kim Novak talked about the scene and enthused about the trees:

> I've always admired trees. I just worship them. Think what trees have witnessed, what history, such as living through the Civil War, yet they still survive. I've always felt that part of why they survive is because they don't try to intercede, to advise 'No, that's the wrong way,' or to try and wipe out an army. They stood and observed.

She then explained that when she first read through the film script and reached 'that part of the Hitchcock script where Madeleine and Scottie are among the redwoods, she touches the tree rings and says, "Here I was born and here I died. It was only a moment. You took no notice," I got goose-bumps. When it came to shoot that scene, I had goose-bumps.' Kim Novak found that 'Just touching that old tree was truly moving to me because when you touch these trees, you have such

93 Kim Novak and James Stewart look up at the sequoias in Muir Woods, California, in a still from *Vertigo* (1958).

a sense of the passage of time, of history. It's like you're touching the essence, the very substance of life.' She remembered taking her father to see the redwoods: 'He wept and so did I. He "got" it in the same way as I do. We never talked about it.'[1]

The scene is set in Muir Woods, named after the well-known naturalist and conservationist John Muir (1838–1914), but it was largely filmed in the Big Basin Redwoods State Park near Santa Cruz, which had been protected through the energetic activities of a much less famous conservationist. Andrew Hill (1853–1922) was a painter and photographer living in San Jose who was also a redwood enthusiast. He was instrumental in saving the large grove of redwoods at Big Basin and getting them taken into public ownership. The Big Basin Redwoods State Park was established in 1902 after a short but intense campaign. The story goes that in 1899 Hill was commissioned by the English *Wide World Magazine* to provide some images of redwoods after a forest fire. He 'photographed the coast redwood trees in Felton Grove' but the owner of the forest 'accused Hill of trespassing and demanded his negatives. Hill refused and left, vowing to himself to save the trees for future generations' and to establish a public park. In 1900 he and a group of friends formed a pressure group named the Sempervirens Club, which held regular visits to the trees and which 'pushed the state legislature to approve a bill for purchase of the land' (illus. 95). This bill also established

94 This annotated cross-section of a redwood (*Sequoia sempervirens*), felled in
1930, near the entrance to Muir Woods, photographed in around 1948, appeared
in *Vertigo* (1958).

the California Redwood Park Commission, which was able to purchase
and receive gifts of land from private individuals and companies. The *San
Jose Mercury* celebrated the success of the park and emphasized the
threat of total destruction that had been averted:

> Giant redwoods. Mighty with the strength which had
> withstood the ravages of centuries, quivered at the menacing
> snarl of the saw mill; trembled with the throb of its engines;
> moaned with the scream of the ripping, tearing steel teeth,
> cutting through the heart of the forest, nearer and even
> nearer and from their towering height beckoned across
> the mountains for rescue.[2]

The Commission's major acquisitions included almost 4,000 acres from
the Big Basin Lumber Company in 1906 and the state park currently
covers over 18,000 acres.[3]

John Muir (1838–1914) was born in Dunbar, Scotland, and moved
to Wisconsin with his family in 1849. He was strongly influenced by
the 'value-centred views of nature' in the works of Wordsworth, Ralph
Waldo Emerson and Henry David Thoreau and by the writings of
Alexander von Humboldt. He first visited the Yosemite Valley, about 200
miles east of San Francisco, and the Mariposa Grove of giant redwoods

(*S. giganteum*) in 1868 and lived there from 1869 to 1871; it became his spiritual home.[4] The grove had been first reported at a meeting of the California Academy of Science by Augustus Dowd, a hunter, in 1852. The English plant collector William Lobb attended the meeting and was so excited by what he heard that he 'raced back to find the trees' where he was 'stunned by their size'. He recorded 80 to 90 trees that were 250 to 320 feet high and between 10 and 20 feet in diameter. He 'collected all the seeds, botanical specimens, vegetative shoots, and seedlings he could carry back to San Francisco' and immediately booked the first passage home to London. Within six months the nurseryman Veitch was offering young trees for sale at £3 2s each and they immediately became an enormously fashionable tree species for garden, park and arboretum.[5] The naming of the tree was problematic. *S. sempervirens* was called Wellingtonia after the Duke of Wellington, who had died in September 1852, by John Lindley in Britain, but the botanist Albert Kellogg in California called it Washingtonia on the basis that a Californian tree should not bear the name of a British soldier and prime minister. This created an international botanical controversy and causes confusion to this day.[6]

While young sequoias were beginning to become established in British gardens and parks, they were photographed at Yosemite in the summer of 1861 by the young American photographer Carleton Watkins (1829–1916), who had established a studio in San Francisco in the 1850s.

95 The Sempervirens Club at the Father of the Forest tree, 1901.

96 Carleton Watkins, *Grizzly Giant*, Mariposa Grove, Yosemite, 1861, photograph.

97 Albert Bierstadt, *Grizzly Giant Sequoia, Mariposa Grove, c.* 1872, oil on paper mounted on board.

He frequently made use of a stereo camera, which allowed people to visualize scenes in three dimensions with the use of a stereoscope, and also developed a very large camera that allowed him to capture images on mammoth plates and produce photographs of great detail and depth. His Yosemite photographs were exhibited in New York in 1862 and people 'around the country were entranced' by them, while 'Ralph Waldo Emerson declared that his images of the massive sequoia, Grizzly Giant, "made the tree possible" for these photographs provided evidence of its existence'. The photographs encouraged the artist Albert Bierstadt to travel to Yosemite and paint the trees and landscapes, and most importantly a set of the prints was owned by Senator John Conness of California, who 'laid the foundations for the Yosemite Bill of 1864 to protect the area from development and commercial exploitation'.[7] Although Watkins's later life from the 1890s was clouded by vertigo and partial blindness and his studio was destroyed in the San Francisco earthquake of 1906, his photographs were much celebrated during his lifetime and were enormously influential in forming knowledge and understanding of the ancient trees and in their protection from logging and development.

Yosemite was made a state park through a federal grant by Abraham
Lincoln in 1864 and then a National Park in 1890. Frederick Law Olmsted
visited Yosemite and wrote a report to the park commissioners in the mid-
1860s emphasizing 'the powerful effect of its picturesque scenery, with
its beautiful fields and groves on the valley floor, giant redwoods, and
sublime granite precipices'.[8] Donald Worster argues that 'The purpose
of parks was to stimulate both the mind and emotions in a positive way,
invigorating the whole person.' They were 'to advance the cause of democratic civilization, and in that movement Yosemite became, after Central
Park, the next step forward'.[9] While John Muir was living in Yosemite,
Ralph Waldo Emerson visited and agreed to view the sequoias with him.
Muir later recollected that as they 'rode through the magnificent forests'
he kept calling Emerson's attention to the sugar pines, quoting his 'Wood-
notes' poems, '"Come listen what the pine tree saith," etc.' Muir tried
to persuade Emerson to camp under the sequoias overnight, but Emerson's
party felt that 'he might take cold' and preferred to stay in the small hotel
with its 'carpet dust and unknowable reeks', which Muir felt to be a 'sad
commentary on culture and the glorious transcendentalism'.

The following morning they rode to the sequoia grove 'and stayed
an hour or two, mostly in ordinary tourist fashion', looked 'at the biggest
giants, measuring them with a tape line' and rode 'through prostrate fire-
bored trunks'. Emerson 'was alone occasionally, sauntering about as if
under a spell. As we walked through a fine group, he quoted, "There were
giants in those days," recognizing the antiquity of the race.'[10] Emerson
was asked to name a tree and 'I selected a Sequoia Gigantea . . . & named
it *Samoset*, in memory of the first Indian ally of the Plymouth Colony.'
He gave instructions for Galen Clark, the forest guardian, 'to procure a
tin plate, & have the inscription painted thereon in the usual form of
the named trees; Samoset 12 May 1871 & paid him its cost. The tree was
a strong healthy one, girth at 2.5 feet from the ground, 50 feet.'[11] When
Emerson and his party left the Grove, Muir was left by himself and he
'sauntered back into the heart of the grove, made a bed of sequoia plumes
and ferns by the side of a stream, gathered a store of firewood, and then
walked about until sundown'. He then 'built a great fire' and was 'lone-
some for the first time in these forests'.[12] Muir left the Yosemite Valley
that year and his later writings, such as *God's First Temples: How Shall
We Preserve Our Forests?* (1876), and many others on the formation and
extension of national parks were enormously influential. He became first
president of the Sierra Club, which was established in 1892 to preserve

forests and other natural features of the Sierra and make them more accessible to the public.

Muir Woods are on the northern side of Golden Gate Bridge northwest of San Francisco near Mount Tamalpais and were designated a National Monument by Theodore Roosevelt in 1908. They contain one of the most famous and well-visited groves of old-growth *Sequoia sempervirens* and their popularity is enhanced by being only 8 miles to the north of the city. The area had become increasingly popular with hunters and hikers from the 1880s and associations such as the Tamalpais Club were established. Some were 'organized by Austrian and German residents who sought to continue a favorite pastime from their native countries, and who likened the scenery of Mount Tamalpais to the Alps'. Trails which took in the redwood groves were built and a scenic railway took tourists to the summit of the mountain. The Tavern of Tamalpais was at the railway terminus; it has 'a long porch facing south, overlooking the Redwood Canyon and the Pacific Coast, with San Francisco in the distance'. Trails led from the tavern down through the Sequoia or Redwood Canyon.[13]

The area became increasingly popular for camping holidays, while hunting continued in the winter months. Many different groups pitched summer camps, including a Presbyterian Church Sunday School Athletic League and the San Francisco Bohemian Club. This had been established in 1872 as an elite social club for gentlemen interested professionally in art, music and drama and by the 1880s was 'one of the most prominent social organizations for wealthy businessmen' in the city. The club held an annual summer camp, usually in redwood groves. By the 1890s this lasted a week and 'regular entertainment involved games and theatrical events, often in an atmosphere of mystery and intrigue.' In 1892 they camped at the foot of Mount Tamalpais and their theatrical high jinks included the construction of a full-scale replica of the Great Buddha of Kamakra for their play 'Bohemia's Redwood Temple' and the holding of a 'Ceremony of the Cremation of Care'. The plaster Buddha collapsed after a year or so, but the road they built to their redwood camp improved access dramatically and allowed many additional tourists to visit the trees, picnic and camp overnight. Perhaps influenced by the Bohemian Club, 'a group of prominent writers from San Francisco', including Jack London, 'chose Redwood Canyon as the spot to dedicate a memorial to the one-hundredth anniversary of Ralph Waldo Emerson's birth'. A small bronze plaque with Emerson's date of birth was attached to what was thought to be the largest tree and during the ceremony a message from John Muir was read.[14] But while tourism increased public awareness and enjoyment of the redwoods, there was at the same time an increasing threat of housing development and logging.

As at Big Basin, it took a committed local man to ensure the preservation of the woodland. William Kent was the son of a wealthy Chicago meatpacker who had established Kentfield, a house and farm of 850 acres, in 1871 for use in the summer. The family gradually bought more and more land and after his father's death in 1901 William Kent became one of the largest landowners in the area and made Kentfield his main home. In 1901 he helped to set up the Tamalpais Forestry Association, which was mainly concerned with protecting the area from fire. He chaired a meeting of the association in 1903 which made a formal proposal for the establishment of a 12,000-acre public park on the mountain, but the Tamalpais National Park Association established at this meeting was not able to gain sufficient funds to be effective. Various key conservationists such as John Muir and Charles Sargent, Director of the Arnold Arboretum at Boston, visited the Redwood Canyon in 1904, but it was William Kent, making use of his local business connections, who eventually purchased the Redwood Grove in 1905 with the intention of opening it as a public park free of charge. He improved the trails and introduced rustic-style seats, tables and log cabins derived from the Picturesque style popularized in America by Andrew Jackson Downing and developed by Olmsted and Calvert Vaux. An extension of the scenic railway 'descended into the woods through a narrow clearing carefully cut through the forest' but was kept well away from the big trees. The Redwood Canyon had by 1907 in effect become a public park, which together with the scenic railway to Mount Tamalpais was one of the most popular tourist sites in the region (illus. 98, 99)

But a new threat to the redwoods very soon emerged. The earthquake of April 1906 caused a great demand for timber and water, and in 1907 the North Coast Water Company, which had the water rights to Kent's property, planned to make a reservoir which would flood 47 acres of the Redwood Canyon floor and destroy many of the biggest trees. Kent sprang into action and cabled Gifford Pinochet, head of the Forest Service, asking for the woodland to be accepted as a National Forest. If the area was made federal land, it would be protected from the activities of the water company. Kent also contacted the forester Frederick E. Olmsted, who advised him, however, that National Forest designation

98 Redwoods in Muir
Woods, California,
c. 1930.

99 Redwoods
at Killerton,
Devon, 2013.

might not preserve the trees, since 'the Forest Reserves policy of 1905 stressed the importance of "use" in National Forests, which was typically understood at the time to mean sustainable timber production.' Olmsted recommended instead that the redwoods could be identified as of scientific interest and hence could be made a National Monument under the Antiquities Act of 1906. This Act gave the president power to designate federal lands National Monuments to preserve resources of prehistoric, historic or scientific interest. It was decided to name the redwoods the 'Muir National Monument', and Olmsted wrote that it was 'of extraordinary scientific interest because of the primeval and virgin character of the forest and the age and size of its trees'. He thought that it 'may some day be one of the very few vestiges of an ancient giant forest' and emphasized that it was 'a living National Monument, than which nothing could be more typically American'.[15]

Very rapidly this plan was put into action. Kent gave the land and trees to the federal government and in January 1908 Theodore Roosevelt signed the proclamation making the woods a National Monument and the first given by a private individual. Roosevelt had suggested that the woods be called the Kent Woods, but Kent preferred to celebrate the conservationist John Muir, who was delighted by the accolade. He wrote to Kent that 'This is the best tree-lover's monument that could possibly be found in all the forests of the world' and joked: 'That so fine and divine a thing should have come out of money made in Chicago! Who wad'a thocht it! Immortal Sequoia life to you.' Kent celebrated the redwoods in an article in the *Sierra Club Bulletin* of June 1908. He admired how from the hillside 'the forest shows a rich and varied coloring. The ruddy tinge of the redwood foliage makes sharp the brighter green of Douglas fir, while softening all is the silver gray of mountain oak.' Within the grove the redwoods have 'thick soft warm tinted bark' with 'delicate foliage' which 'sifts the sunlight, not precluded, but made gentle'. Kent moralizes that the redwoods are 'brave' and 'Burned of all their leaves, they fight for life and bourgeon [sic] out again. Around the fallen parent grows up a stately group of children.' He concludes by predicting that an 'American Wordsworth will one day sing these noble trees as teaching the ideal of the social and individual life of the Americans.'[16]

National Forest Parks in Britain

The concept of national parks spread widely in the late nineteenth century and they were designated in Australia (1879), Canada (1885) and New Zealand (1887), but in a densely settled, intensively cultivated and heavily industrialized country like Britain it took a long time for the idea to take hold. There were no large areas of what could be seen as pristine landscapes that needed to be preserved, and anyway all the land was owned by someone, and private owners were not at all keen to see their development rights restricted by conservation designations. But by 1929 the pressure to improve access to the countryside was mounting and the Addison Committee was set up to examine the feasibility of national parks and of improving public recreation in the countryside.[17] Many landowners, including the Forestry Commission, which had been established ten years before in 1919, were worried about the effects of untrammelled public access. The Forestry Commission noted the potential conflict between increased public access and the preservation of flora and fauna. It was felt, for example, that 'the erection of Hutment camps around the Forest of Dean might tend to destroy the amenities of the neighbourhood and the process might be completed by ill-disciplined visitors, charabanc parties etc.' while 'the institution of a camp for children might imperil a Bird Sanctuary, and so on.'[18] This internal Forestry Commission mimeograph went on to report Lord Bledisloe's ideas for national parks, 'where people of all walks of life can enjoy under proper protection and with reasonable comfort attractive natural surroundings'. Bledisloe, an influential landowner and agricultural reformer who became Governor General of New Zealand in 1935, also proposed that the parks should include 'permanent camps provided with water and sanitation, refreshment and entertainment, bungalows, car parks, and, if possible, open air swimming baths, bowling greens, tennis courts etc.' This seems very distant indeed from the idea of national parks espoused by John Muir at Yosemite.

The National Parks Committee reported in 1931 and supported the idea of national parks, but no mechanism for their establishment was introduced. The idea continued to be debated, however, and in 1934 Peter Thomsen, in a paper to the British Association for the Advancement of Science at Aberdeen, suggested that there should be a threefold approach to establishing national parks. First, areas that were to become national parks should be scheduled and in those areas 'development' should be 'vetoed'. This was to include the prohibition of 'the cutting of timber, especially old timber'. Second, limited access in the form of 'rest houses' and 'camp sites' should be established; and finally, the government would arrange 'full possession upon purchase' of the national parks. Thomsen's ideas were similar to those put in practice at Yosemite and other American national parks. A copy of Thomsen's paper was read by Sir George Courthope, who sent a copy to Sir Roy Robinson, chairman (1932–52) of the Forestry Commission, in October 1934, noting 'Have you seen the enclosed effusion about National Parks? I am sure that you agree that wild schemes of this kind must be nipped in the bud. With this end in view, do you think it would be a good thing to push forward your proposals for recreational facilities in connexion with our forests?[19] George Loyd Courthope (1877–1955) was an influential landowner and spoke for the Forestry Commissioners in Parliament. His comments demonstrate the close organizational links between private landowners and the Commission, and also show the extent to which both the state-run Commission and private landowners felt threatened by the possible establishment of national parks, which could limit large-scale afforestation and inhibit the management of established woodland for commercial purposes. The Forestry Commissioners were not opposed to the idea of public access, but they and the landowners who supported them were very much opposed to the formation of national parks run by bodies that would interfere with traditional land management and imperil large-scale afforestation of moorland.[20]

Almost immediately after Sir George Courthope's suggestion about 'pushing forward' with 'recreational facilities', the Forestry Commissioners established a National Forest Park Committee 'to advise how the surplus and unplantable land' on Forestry Commission property in Argyll might be 'put to a use of a public character'. Their 1935 report described how discussions had been held with youth hostelling and other associations and societies and they felt that the hundred square miles of largely unplantable moorland in Argyll was ideal for 'the rambler, whose main object is to get into the country and away from motor traffic'. Responsible organizations such as the Scottish Association of Boys' Clubs and the Camping Club of Great Britain would enforce by-laws so that 'decent behaviour' could be encouraged and it was noted that the risk of damage could be reduced by the careful design of paths, which should 'pass to the uplands through the afforested lands by easy gradients.

This would have the effect of keeping people from trespassing on or causing damage to plantations.' The report cautiously recommended the establishment of a National Forest Park and that the government should provide £5,000 for this purpose, although 'campers should in every case be charged a fee to cover the expenses of the services provided.' The wilderness quality of the Argyll Forest Park was stressed and the designation was certainly seen as experimental. It was 'difficult to estimate how many people could make use of the area . . . for recreation, without destroying the sense of remoteness and solitude which is its chief attraction' and the committee rather nervously recommended that 'the Commissioners should proceed cautiously and refrain from drawing undue public attention to what they are doing' but it did hope that 'experience gained here should be of use in other areas belonging to the Commission.' The Committee was keen to establish that the term National Forest Park was 'deliberately intended to denote something different from a National Park as described in the Report of the National Park Committee' of 1931.[21]

Sir Roy Robinson argued in 1936 that although the Forestry Commissioners were 'thoroughly sympathetic to the idea underlying National Parks in general' they considered that 'proposals to sterilise large tracts of country and especially . . . to ban plantations on suitable ground in National Parks should be scrutinised very carefully.' He thought that Britain was too small for the large national parks found in 'America and the newer countries generally' and defended afforestation by asserting that plantations might well be an asset in 'wild country', providing 'shelter for the wayfarer' as well as 'constituting in times of emergency an essential raw material'. He described the Argyll National Forest Park as a potential forerunner of a different model of public access and as 'something of an experiment'. The Commissioners knew from 'their management of the New Forest, which is the nearest approach to a National Park that we have, what are the difficulties in reconciling the protection of forests with access for the public', but they believed that 'with care the two can be reconciled.' Robinson saw forest parks as a 'by-product' of the Forestry Commission's primary function of timber production, but a by-product 'which it may well pay the country to develop'. The Commission already held 333,000 acres of unplantable land and was 'constantly acquiring land in various parts of Great Britain', and if Parliament thought it desirable it could focus more on purchasing land suitable for national forest park purposes.[22]

The national park movement was gaining strength in the late 1930s. A Joint Standing Committee for National Parks was established in 1936 and soon became a very effective organization with broad-based support from the Council for the Preservation of Rural England, the Friends of the Lake District and the Ramblers' Association. But the Forestry Commission was cleverly positioning itself to scupper the threat posed by the proposed national parks to its freedom to afforest moorlands. In the debate on national parks in the House of Commons of December 1936, the main government speaker, R. S. Hudson, parliamentary secretary to the Ministry of Health, saw the Forestry Commission's Argyll National Forest Park as 'a useful way out' of the problem of establishing 'national reserves'. Moreover he used the existence of the Argyll National Forest Park as a means of side-stepping the issue of establishing a new authority to oversee the formation of national parks. The government's view continued to be that national parks or reserves should be the responsibility of local authorities and not central government. Sir George Courthope's scheme to nip the idea of national parks in the bud had succeeded, at least in the short term.[23] Of course, national parks did eventually come into existence after the war with the National Parks and Access to the Countryside Act of 1949, but a positive outcome of the Forestry Commission's 1930s skirmish with the idea was that it was decided in 1937 to establish a Snowdonia National Forest Park, and the next year (1938) a national forest park was proposed for the Forest of Dean.

National Forest Park Guides

A series of seven substantial illustrated national forest park guidebooks was produced by the Forestry Commission. The first was published in 1938 for the Argyll National Forest Park and after the war came one for the Forest of Dean (1947) closely followed by that for Snowdonia (1948). They continued to be published until the 1970s. The idea of the booklets was raised at the second meeting of the National Forest Parks Advisory Committee (England and Wales) on 17 July 1939, and although there was little discussion concerning the level of popularity of the guides, there was general agreement that the booklets should be informative and educational. Each had a similar format, with cultural chapters on the history and antiquities of the area and on literary associations; topographical chapters on geology, mountains and rivers; naturalist chapters on plant life, mammals, birds and insects; and a chapter dealing with forestry and woodland management. Finally there are maps showing the extent of the park and the footpaths and information about camping facilities. Most of the guides, other than the first, were edited by Herbert Edlin, who was Publications Officer for the Forestry Commission from 1947 and a prolific author. His *British Woodland Trees* of 1944, published by Batsford, was enormously successful and was soon followed by *Forestry and Woodland Life* (1947), *Woodland Crafts in Britain* (1948) and many other books.[24] Edlin argued that the forest parks had an advantage over national parks in that they were wholly owned by the state, whereas national parks were privately owned.[25]

The forest parks and the guides that were produced so carefully to inform and educate the public were an attempt by the Forestry Commission to interest the public in a reformed and modernized forestry that was to transform the uplands of Britain. In 1950 the Commissioners worried that while they had 'consistently striven to keep the public informed of their objectives and of the progress of their forest operations, they have found it no easy matter to get the facts across to the man in the street'.[26] The guides were part of a strategy that also included special forestry broadcasts on the BBC, the development of educational material for schools and the display of scale models of the Snowdonia National Forest Park alongside scale models of modern forestry practice at the Festival of Britain exhibition in 1951.[27]

How was recreation to be encouraged and managed in these new forest parks? The guides celebrate the parks as places for the efficient production of timber using the most modern methods, yet also as containing large remnants of wilderness for the hiker. The parks are represented as habitats for flora and fauna for the serious naturalist and delightful and diverse landscapes for the casual visitor. In some ways the guides fell between two stools: they were too dry or scholarly for the casual visitor and not detailed enough for the serious walker. There was a tension between the need for the Forestry Commission to retain very close control of the public on Commission land and yet at the same time promote the idea of the forest as a wilderness to be freely explored by a public who understood the workings of the countryside. There was concern that access was not easily reconcilable with the establishment and maintenance of young plantations of trees: fences could be damaged, trees stolen or uprooted and the risk of fire increased. But it was also agreed that the public should be allowed to roam fairly freely and not cajoled into following prescribed routes and paths. The Commissioners proposed at Argyll Forest Park in 1935 that feasible routes should be marked 'by occasional cairns or whitened stones, or distinctive waysigns or symbols'.[28]

People were to be enticed to the new forest parks by the provision of bureaucratically delineated types of camping facility. At the Forest of Dean in 1938 it was envisaged that the park would provide three types for visitors: 'Huts and chalets, with dining and recreation hut and the pro-vision of meals'; 'Tents to be supplied at a reasonable charge, campers to provide their own meals' and 'Sites for visitors carrying their own equip-ment and providing their own meals'. But the Commission was reluctant to provide any accommodation in the forest parks if, as in the case of Snowdonia, it was felt that there was enough private accommodation available already in the form of youth hostels, hotels and bed and break-fast facilities. Different parks had rather different priorities. In contrast to the Argyll Forest Park, day trippers were encouraged at the Forest of Dean by providing 'pull-ins' for motorists, and people were to be encouraged away from the beaten track and out into the woods by the construction of footpaths. It was recognized that the woods along the Wye Valley

> are on hilly land and the paths are rough and, generally speaking, the public do not wander far into them . . . These valleys, beautiful in themselves, lead to high ground with an elevation of as much as 1,000 feet and we think that the public will, in the course of time, wish to explore these places and that consequently improved access will be necessary.

Yet in spite of the fine words, it is clear that the Forestry Commission was very reluctant to commit itself to any but the most minimal of facilities. This approach could be justified both on the grounds of cost, because of the stringencies of post-war austerity, and as a way of limiting the environ-mental intrusion of tourist facilities.[29]

By the early 1960s the Forestry Commission's access policy and particularly its guidebook literature were to come in for severe criticism. A mismatch was identified between 'the interests of the forest manager, which commonly may lean towards natural history' and that of the general public, and it was noted that the 'lengthy and descriptive' guides with their 'botanical, zoological and archaeological' content were only 'absorbingly interesting' for 'the fairly small proportion of walking-holiday visitors', while for the majority of visitors who were 'car driving, day-visiting family parties' they 'make an insignificant impact'.[30] The guides did have a role, however, in popularizing coniferous aesthetics and public access to the Forestry Commission's estate. One of the difficulties Edlin had in publicizing forest parks was that the name conjured up so many contrasting meanings: hunting forests; commercial timber production; freedom to roam; and controlled access. Moreover although some of the forest parks, especially the Forest of Dean, had well-established areas of broadleaved woodland, the majority consisted largely of thousands of acres of young conifers, which most people did not find immediately attractive. In his review of 50 years of national forest parks, Edlin wrote of Kielder Forest in Northumberland that 'Spruce woods on soggy peat, stretching over rounded hills that are often misty or cloud-capped, are not ideal for outdoor enjoyment; and the fact that they go on for further than anyone can walk is more daunting than encouraging.'[31]

The massive use of coniferous trees by the Forestry Commission resulted in new large-scale landscapes for which there was little precedent. William Wordsworth in 1835 was famously critical of the effect of larch plantations on the landscape of the Lake District:

> a moment's thought will show that, if ten thousand of this
> spiky tree, the larch, are stuck in at once upon the side of a
> hill, they can grow up into nothing but deformity; that, while
> they are suffered to stand, we shall look in vain for any of
> those appearances which are the chief sources of beauty in
> a natural wood.[32]

Between 1920 and 1938 the area of land under forest crops held by the Forestry Commission increased from 1,393,000 to 400,712 acres. The great bulk of this afforestation was coniferous and Miles Hadfield notes that during the interwar period 'a sense of urgency and enthusiasm unusual in any government-controlled body inevitably led to many mistakes and plantings that were both unsightly and unsatisfactory, both by the Commission and the landowners it assisted.'[33] The Forestry Commission's interwar afforestation of open moorland and heath resulted in industrial, modern, regular, efficient and utilitarian landscapes that soon began to be widely criticized. The two main arguments against these industrial-scale landscapes were that there was a loss of public access over open moorland which had been afforested and that there was a change from a 'wilderness' landscape to one that was obviously managed and productive. The pre-war controversy over afforestation reached a peak with proposals to afforest parts of the Lake District. It was claimed that 'public access to the open fells is endangered' and that the proposals valued 'the profits of commercial timber more than health and beauty'. The Forestry Commissioners eventually reached a compromise with the Council for the Preservation of Rural England (CPRE) and undertook not to plant up the central section of the Lake District.[34]

The visual intrusion of coniferous plantations was never far from the thoughts of Forestry Commission publicity. As late as 1969, well over 30 years after the Lake District controversy, Edlin was pleased to observe that many visitors enjoyed walking in the plantations of Snowdonia, whose

> forests are surrounded by the grandest mountains of England
> and Wales, and stand close to long sandy beaches, yet on
> every fine summer's day they are filled with visitors who
> could go elsewhere, but prefer to seek the peaceful fascinations
> of growing timber crops. The prophecies of critics who declared
> that people would shun the 'dark, dreary, dismal conifers' have
> been confounded.

The national forest park guides, therefore, had the difficult job of encouraging public appreciation of huge new afforestation schemes. This was particularly difficult where there were few older plantations. At Gwydyr Forest in Snowdonia the woods 'clothe the side of steep valleys and extend over rugged foothills studded with still lakes. The whole has been steadily afforested, during the ensuing half century, with plantations of larch, pine, spruce and Douglas fir that now look entirely natural'. But this was not a common state of affairs in the 1950s, when most Forestry Commission plantations remained young and often monotonously featureless.[35]

One of the ways Edlin dealt with this in the guides was in the very careful choice of illustrations. These included photographs but also line drawings and most characteristically a series of woodcuts and wood engravings. By the mid-1950s all the guides had cover illustrations by George Mackley, who had been taught by Noel Rooke, one of the 'chief originators of the modern movement of wood-engraving', who helped to 'reinstate the "white line" technique that Thomas Bewick and William Blake had developed some hundred years earlier'.[36] Given the hostility experienced by the Forestry Commission over the visual qualities of its young plantations and the absence of mature plantations to photograph, it is perhaps not surprising that it should resort to 'artistic impressions' of mature forested land in order to sell the forest parks to visitors. But this was an attempt not only to ameliorate the landscape of commercial forestry, but to popularize an alternative large-scale coniferous aesthetic, as in the Argyll Forest Park, where a sense of 'forest' was created by 'thousands of acres of timber crops'.

'In the Sprucewoods' is the frontispiece to the forest guide to the Border Park, the seventh forest park, which had been opened in 1955 and included the vast Kielder Forest. This guide was edited by the botanist Professor John Walton, who emphasized that the Border Park was notable for containing 'the largest planted forest in Britain', which would 'supply the nation with a substantial source of timber' and was planted on 'hill land of low agricultural value which supported but few grazing animals'. The afforestation had led 'to an increase in the population of the district, an increase which will be progressive as the forests mature and local crafts and industries relating to forestry develop'. He foresaw this 'sparsely populated land developing into an active and prosperous rural area'.[37] Only a few years were to pass before this engaging and convincing argument for afforestation was shown to be spurious, as the employment generated by forestry started to fall dramatically. The predominant tree species planted in these forests was the Sitka spruce (*Picea sitchensis*), which grows naturally in the coastal regions from Alaska down to northern California. It was first recorded by Archibald Menzies at Pujet Sound in 1787 and seed was sent to the Horticultural Society of London by David Douglas in 1831. In Britain it was recognized that it had the advantage of growing well on poor upland soils and a 'period of major afforestation occurred from 1950 until the late 1980s', which resulted in the area of Sitka spruce increasing from 67,000 ha (165,560 acres) in 1947 to 692,000 ha (1,710,000 acres) in 2007.[38] One of the jobs of the Border Park guide was to accommodate this massive change in the eyes of the general public.

George Mackley's *In the Sprucewoods* is a celebration of coniferous forestry. The woodcut is a particularly appropriate means of representing living timber and this is emphasized by the specification in this guide that this was a 'boxwood engraving'. The adoption of woodcuts in

100 George Mackley, *In the Sprucewoods,* 1958, wood engraving.

these guides helped to naturalize the 'exotic' and 'alien' forests of Sitka
spruce. The picture shows a range of different aged stands surrounding
a stone farmstead and mill besides a rustic arched bridge. The detail of
the woodcut emphasizes the textural variety of the scene in a conven-
tionally Picturesque formulation. There are a few broadleaved trees

101 George Mackley, cover for *Border National Forest Park Guide* (1958)
wood engraving.

growing behind the buildings, but the image is dominated by vigorous and rapidly growing conifers whose darkness threatens to overpower the lightly coloured buildings. The young plantation on the hill indicates a landscape newly transformed by afforestation, with a small patch of walled pasture remaining. Mackley's wood engraving for the cover is of a view through a working forest across to afforested hills and moorland. Some Sitka spruce have already been felled and are being moved by horses and stacked for removal. The representation of the horses and the forester with an axe portray methods of working that were very rapidly to disappear with mechanization. The image became anachronistic within five to ten years, but we can see it as a brave attempt to show in a most advantageous light a landscape which Edlin in his more reflective moments could only see as less than attractive for the visitor. But the guides were also a way of envisaging the future and Edlin thought that it was 'fair to ask the acutely aesthetic minded not to take the short view, but to look ahead to the mature forests of future years, to a forest industry and the busy and increasing community of forest workers who have already begun to find a home and livelihood within afforested area'. This, he argued, could be combined with the benefits to the 'welfare of the generality of the people of Britain', who would be able to 'find healthy recreation in and about the forest and in the National Forest Parks on which a start has been made, as the peoples of other nations are finding it in theirs'.[39]

The guides fit into a broader tradition of the development of educated access to the countryside. In the twentieth century there was a growing interest among many adults in the study of botany, ornithology and general natural history, but there was concern about the behaviour of those people visiting the countryside and how they could learn about nature and understand how to comport themselves in a 'proper' manner. The forest guides combined frequent calls to avoid damage to fences and farm crops, and to guard against accidentally causing fires, which could destroy young plantations, with specialist chapters on botany, ornithology and local antiquities and history. They exemplify the trend towards educated access which was to be strengthened in the 1960s with the spread in Britain of the idea of the nature trail.

The idea of a trail along which individuals could walk and learn about nature from labels, either by themselves or with a guide, was developed in the 1920s by the entomologist Frank Lutz, who worked at the American Museum of Natural History in New York.[40] Forty acres of woodland and meadow in the Ramapo Mountains near Tuxedo, New York, had been granted to the Entomological Department of the Museum of Natural History by the Interstate Palisades Park and W. A. Harriman. This was an area where 'excursion steamers brought thousands of New Yorkers up the Hudson River', many of which were 'children and young people who lived and worked in the city'.[41] A station for the study of insects was established in 1925 and in the first season an 'experiment in teaching about nature in general' was undertaken.[42] Two nature trails were set out, a training trail and a testing trail: plants were labelled and stories told. The experiment was written up and published the following year and was seen as very successful; concerns over damage from visitors were unfounded. Lutz argued that 'National and State parks are particularly good fields for Nature Trails' and that the 'the very best way for making a Nature Trail is . . . to have as many as possible of those whom you wish to teach help you to make it.'[43] It was best to 'teach about nature where nature is' rather than in museums. The publication of the experiment popularized the idea of nature trails; many were established in national parks and in 1930 'much interest' was 'shown in the new nature trails at Yosemite National Park'.[44] The nature trail, like the national park, was an idea that soon spread to Britain and woodland nature trails became commonplace from the 1960s onwards.[45] By the end of the twentieth century many woods and forests in Britain were celebrated and valued more as sites of recreation and conservation than as places for timber production.

TEN
Ligurian Semi-natural Woodland

THERE has been a remarkable increase in woodland in many parts of the Italian Apennines since the end of the nineteenth century. If one climbs to one of the many viewpoints, as at Costa dei Ghiffi (illus. 102), one looks over a lush, arboreal landscape only occasionally interrupted by an outcrop of rock, church tower or patch of farmed land. Only 100 years ago such a scene would have been characterized by many extensive, open, grazed areas and heavily cultivated agricultural terraces producing crops of potatoes, maize and vegetables.[1] This dramatic change has been brought about by a number of factors, amongst the most important of which is very extensive rural depopulation, which has left small towns and villages partially moribund and resulted in the abandonment of many hamlets and isolated farms. Linked to this has been the collapse of the traditional transhumance systems used to facilitate the grazing of most upland areas. The massive reduction in agricultural and pastoral pressures has allowed very significant areas of naturally regenerated woodland to develop. Detailed case studies from the upper Val di Vara in eastern Liguria show the complicated relationship between the decline of traditional practices and techniques and the new woodland. Four important issues are the decline of traditional shredding of trees for the production of leaf fodder; the regeneration of trees within former wood pastures; the growth of secondary woodland on abandoned agricultural terraces; and the loss of traditional chestnut cultivation and the neglect of chestnut groves and orchards. The vast new areas of Ligurian semi-natural woodland form a sort of unplanned experiment in which the advantages and disadvantages of allowing the rewilding of landscapes can be considered.

102 Costa dei Ghiffi, 2014.

Leaf fodder in the Val di Vara

In the crypt of the abbey church at Bobbio there is a fascinating mosaic calendar showing the different months of the year dated to the second half of the twelfth century. The image for November shows a man beating an oak tree to knock down acorns for two pigs that are rooting around the base of the tree. The form of the tree, in this area most likely the Turkey oak (*Quercus cerris*), indicates that it has been pollarded several times in the past. Moreover, the stubs around the trunk indicate that the tree has had lateral branches cut off, as in the process of shredding. This is the earliest representation we have of a shredded and pollarded oak tree in the Ligurian Apennines and the fact that this practice was chosen to symbolize farming activity for November is an indication of its prevalence and importance. Bobbio Abbey was founded in 614 by the Irish monk St Columbanus and eventually became one of the most important landowners in northern Italy. A map of 1564 that was used as evidence in a border dispute between Bobbio Abbey and the village of Coli in the Trebbia Valley provides another indication of the importance of trees as sources of fodder.[2] This shows oak trees which have been shredded; they are drawn on the map as individual trees rather than as woodland. Here we have an example of what Diego Moreno and Roberta Cevasco have identified as an agro-sylvo-pastoral system for leaf-fodder production. Sometimes small patches of earth were cultivated and cereals sown between the trees in an intensive system called *terre alberate*. Today there is still evidence of these practices on a small scale in remote and isolated Apennine valleys.

There was a strong link between different types of sheep keeping and different types of leaf fodder production. Large flocks of sheep and goats that were kept in olive grove areas around Levanto and Chiavari on the coast in the winter were moved to high Apennine pastures in the summer. But flocks of sheep were also kept by local people in the Apennines all year round, in buildings, and a distinction was made between these local flocks (*pecore terriere*) and the transhumant flocks (*pecore forestiere*) in the taxation records. The 'peasant-shepherds' in the hamlets and villages around Varese paid the landowners cash or shared the cheese produced.[3]

Since at least the mid-sixteenth century it was common for trans-humant shepherds to feed their flocks with leaves directly from branches

103 Mosaic depicting November in the crypt of the Abbey of San Colombano, Bobbio, 12th century.

cut from trees growing in areas of common woodland during the summer. Leaves were also cut in the summer and dried to make fodder that was used as a supplement for other fodder for sheep that were housed in buildings in mountain villages over the winter. Each of the parishes in Varese had two woodwards (*campari*) who controlled the activities of the peasants and shepherds, and legal records show that there were disputes between private landowners and their neighbours and commoners over the effects of cutting fresh branches and the damage caused by browsing sheep and goats. In the parish of Caranza precise laws over leaf fodder production in common woodland were made to control the pressure of both transhumant flocks and stabled animals. Leaf fodder was reserved specifically for the stabled flocks kept by local villagers and its use by transhumant flocks was expressly forbidden.

In the nineteenth century Turkey oak woodlands were normally 'trimmed for browse in summer on a three- or four-year cycle', and trees in woodland, woodland pastures and isolated trees were regularly managed in this way. A forest inventory of 1820 identified 21 local types of managed woodland or lands bearing trees, all of which were grazed, and sixteen types were managed for the production of leaf fodder. Even in their winter quarters on the Riviera 'an arboreous culture provided leaves for sheep': in the autumn fresh leaves from vines and fig orchards were collected by hand and sold for cash as fodder for sheep well into the nineteenth century. At Moneglia, for example, plantations of fig trees on the slopes facing the sea still existed in 1820 and provided a source of food for around 3,000 sheep. Moreover in the winter, the prunings of leaves from olive trees 'provided a considerable amount of green fodder'.[4]

The transhumance system broke down in the late nineteenth century, but some of the traditional practices were kept up by local farmers who had their own flocks, the successors to those who had held *pecore terriere* in the Middle Ages. A photograph from the late nineteenth century of the small town of Varese Ligure captures in the background some stripped oak trees which grow along an ancient pathway leading from the church through the hillside terraces. These trees were stripped until the early 1990s and even today the trees show evidence of pollarding in the form of the multiple branches growing from the trunk. A photograph taken in 1997 shows an oak tree where the lower branches have been cut and the branches have regrown and nearby one of the characteristic willow pollards that are regularly cut to provide ties for vines. It is clear

104 Panorama of Varese Ligure, a photograph of *c.* 1890.

105 Shredded oak tree, Varese Ligure, 2008.

106 Shredded oak and pollarded willow, Varese Ligure, 1997.

that the terraces had been fairly recently abandoned, and indeed in 2010 both these trees were cleared away with the surrounding terraces to make a new large field for growing sainfoin for cattle fodder.

In 2008 we interviewed two farmers in the village of Teviggio, Val di Vara, where some recently shredded trees had been spotted to try and find out why trees were still shredded and to document the history of the methods used. Both farmers were able to point to particular tree species and give a detailed account of changes in the way they were utilized over the last 50 years or so. Marco had been born in 1940 and had inherited his farm and farmed it all his life. He described how the different tree species could be used for different, sometimes very specific purposes. Alder trees (*Alnus glutinosa*), for example, were shredded until about twenty years ago and the leaves were used to fertilize the soil. He recognized that manna ash trees (*Fraxinus ornus*) produced the best leaves for fodder but noted that there were very few growing on his farm. Another commonly shredded tree for fodder was the hop hornbeam (*Ostrya carpinifolia*) and this had been used frequently in the past.

However, the main species used at Teviggio were Turkey oak and Sessile oak (*Quercus petraea*), because they were very common in the area, much more so than ash or hop hornbeam. The two species of oak were intermixed in the woodland and were treated in the same way. The leaves were cut in August, when it was felt that the leaves were in the best condition, and always at the full moon. This was thought to prevent the rotting of the fodder leaves. Short branches were cut with fresh leaves that were dried for two or three days. The trees were shredded when their branches were longer than 1 metre (3¼ feet). Marco remembered that he shredded trees in group of about twenty trees as it was 'harder to shred scattered trees'. Generally this farmer shredded trees on a two-year cycle: 'I cut different trees in different years. In the third year I went back to the trees which were shredded in the first year.' The advantage of cutting every two years was that if you left it any longer the 'branch was too hard to cut'. The tool used to cut the branches was a little axe called a *piccozzo*; this had 'always been used'.

The leaves were stored in *fuggia*, that is, a pile of between 150 and 200 branches with dried leaves that were placed on a platform raised above the ground so that the leaves did not get damp. On this farm two *fuggia* were needed to store all the bundles. The dried leaves were moved from *fuggia* in outlying woods to the cattle stores by mule, and this farmer was considered lucky by his neighbours to have a mule to undertake this work. The leaves were fed to the stock by tying a bundle of dried leaves to the wall of the stall and the animals used the leaves to supplement their feed from grass hay. Any remaining twigs and branches in the bundle, which by winter had dried out entirely, were used to heat bread ovens, representing a very efficient dual use of the collected branches. The only green tree in the winter was *Erica arborea*, and this was used as winter feed for cattle.

The tops of pollarded trees were also used for fodder, and pollarding produced a lot of useful sprouts. The Forestry Authority did not allow Marco to cut the tops of trees, but, even so, he still cut the tops of some, leaving a few branches to avoid getting into trouble, although he had 'been fined several times which is why he left a few of the top branches'. He admitted that farmers and foresters have different ideas about trees, the farmers being interested in the production of fodder and the authority wanting trees to be tall and beautiful. He noted, 'I understand why the law is in place to produce good timber but I needed to shred the trees to feed my animals.' A local forestry official pointed out that shredding of trees has been illegal since the early nineteenth century. However, shredding was common until 40 or 50 years ago, and he knew that a few farmers still shredded trees occasionally, but 'ignored it'.

The second farmer, Enrico, came from a farming family and had inherited his farm, although for many years he had worked away from the farm. He had been born in 1944 and remembered that until the 1950s shredding was commonplace. His uncle cultivated the farm for maize, potatoes and hay until 1962, when the land was abandoned. This farmer moved back to the farm in 1994, after it had been abandoned for 30 years. He remembered that the leaves of ash, oaks and chestnut were used for fodder. His testimony agreed largely with Marco in that the leaves were always cut with the old moon, otherwise they would rot, and that leaves were cut on a two-year cycle. He also stated that both Turkey and sessile oaks were used for fodder, but thought that cattle preferred the leaves of sessile oak, which is why they are the largest surviving shredded trees in the area. He provided more information on the technique of shredding. You would climb to the top of the tree and start there, cutting off branches as you came down. One or two branches were left uncut at the top 'to keep the tree alive'. The branches were cut a few inches away from the trunk to ensure that there was future growth and prevent the branch from becoming dormant. The two main tools used were the *piccozzo* (as with Marco) and the *roncola*, which he preferred as 'it had a short handle which made it easier to cut quickly.'

The branches were left on the ground near the tree to dry and divided into those suitable for fodder and firewood. It took three or four days for the leaves to dry, ready to be stored, and the branches were turned twice a day as for hay. Enrico thought that the best-quality leaves were those that had dried in the shade rather than in full sun. There was a gender division in the work here, with the men cutting the branches off the trees and sorting the larger branches for firewood, whereas the women dealt with the leaves for fodder and turned them to dry. When dried, the fodder branches were tied together with supple stems of chestnut called *struppelli*, which were twisted apart from one end and wrapped around the bundle to hold it together. The bundles were then stored either in a barn near where the animals were kept or in a *fuggia* made in the woodland. Bundles were taken to feed the animals as and when needed. The last time he remembered shredding a tree in the traditional way was in 1975. He no longer shreds because he is now 'richer and the practice was too much hard work'. Nowadays he prefers to buy in feed for his cattle.

Changes in the farming economy, the introduction of new breeds of cattle with higher milk yields and different feeding requirements and the lack of local labour mean that shredding is now a very rare survival. In Teviggio in 1950 there were seven farms that had 100 cows altogether; now the same amount of land has 25 cattle. The old cows produced 5 litres of milk per day; the current cows produce 30 litres of milk. The new cows only eat hay fodder, in addition to some grazing and concentrates. Comprehensive shredding stopped in about 1960 and nowadays, the principal use of wood is firewood. Marco coppices quite a lot of this, both for the wood produced and to increase the amount of land useful for grazing. In the past the woodland areas were grazed by domestic animals and patches of land were cultivated for potatoes, maize and wheat. In addition some patches of rye were sown until around 1950. Branches of *Castanea sativa* were used as brooms to sweep aside unwanted branches and leaves of Turkey oak; this allowed the ground area to be more easily grazed. Another reason for managing the woodland today is to improve the quality and quantity of the porcini and other mushrooms that grow in the Turkey-oak woodland. The idea is to manage the oak to promote its growth and increase the amount of *funghi di cerro*. The lower branches are cut to allow more sunlight to reach the herb layer. The value of the mushrooms is indicated by the need now to fence these woodlands against the roads so that people are discouraged from stealing them. They have a high value and some eventually reach the American market.

Woodland succession in oak-wood pastures

Canavadigiolo is the name of a farm in the parish of Teviggio which, like most parishes within the Alta Val di Vara, has experienced very considerable population loss over the last 100 years. Archival evidence from the Archivio de Paoli held by the Museo Contadino at Cassego indicates that Canavadigiolo, like several of the adjoining farms, had once been part of the Porciorasco estate. Records from the 1820s and '30s indicate that the main products of the farm included wheat, beans, chestnuts, grapes and cheese.[5] Oral history and observation indicates that there had been considerable investment in the late eighteenth and nineteenth centuries in the construction of farm houses, buildings and water-supply channels. The terraces at Canavadigiolo were used to grow vines in the late eighteenth and nineteenth centuries. The archives list the weight of grapes owed by tenants of Canavadigiolo to the estate from 1798 to 1863.[6] After that date there are no records, but in 2003 there was clear field evidence of remnant old vines surviving, so it is probable that vine production continued on the farm at a small scale through most of the twentieth century. The owner informed us that the farm had been gradually abandoned from the 1960s onwards. Until 1952 the chestnut woodland had been partially irrigated from water channels. The land had been worked and cultivated until 1962 and the oak trees continued to be shredded through the 1960s, with the last trees shredded in 1972. Some minor management may have taken place in the 1970s, but by the late 1970s the land was completely abandoned until the owner started working the farm again in 1997.

The area is generally heavily wooded with some abandoned terraces and several completely abandoned farms. I had examined with Diego Moreno in March 2002 an area of what appeared to be abandoned wood pasture, and in particular a stand of Turkey oak that had been shredded at some time in the past. The owner told us that many of the trees had been cut for fodder on a regular basis. A preliminary examination showed that many of the trees were hollow and probably rather older than their appearance suggested. We cored ten trees and the number of rings counted varied from 32 to 220+ years. Two of the trees had high ring counts, indicating that they were much older than had been estimated by evidence from oral history. One tree showed several periods of very narrow ring growth over 20- to 25-year periods, which could be associated with regular pollarding or lopping over a 200-year period. Following the success of this trial 50 trees were cored in the summer of 2002.[7] The number of rings varied from 20 to 240. Many of these trees were hollow and so these are minimum ages for the trees. These results indicate that shredded Turkey oaks have been an important part of the vegetation of the site for over 200 years and probably much longer.

Since the area was abandoned in the 1970s many young saplings and shrubs had spread into the formerly open areas between the trees. We surveyed this vegetation in 2008 and found that overall, hop hornbeam and Turkey oak were ubiquitous while juniper, manna ash and hawthorn had a more localized distribution. The average age of the trees and shrubs sampled was seventeen years, the hop hornbeam averaged 23 years and the Turkey oak nineteen years. Overall the survey showed that there was fairly rapid regeneration of trees and shrubs following the abandonment of the farm in the 1970s. The oral history evidence and vegetation evidence fit together well, and suggest that until the 1950s this area consisted of an area of standard Turkey oak and chestnut trees that were cropped for leaves, nuts and possibly acorns. The details of the management of the pasture under the trees remain unclear. There was some fragmentary evidence of small pocket terraces around some of the older trees, which will have helped to maintain the productivity for chestnuts. It is possible that some areas may have been cultivated on a temporary basis for grain in the nineteenth century and earlier.

With the gradual reduction in management in the 1960s and almost complete cessation of grazing in the 1970s, the natural regeneration of species such as tree heather, juniper, hop hornbeam and manna ash was able to take place. This meant that the gaps between the old pollarded and shredded trees were infilled with secondary growth. Interestingly, some of the tree heather had multiple stems, indicating that although the age of individual stems might be twenty or thirty years old, the age of the stool might be much greater. Oral evidence suggested that the wood was valued as firewood and possibly the making of pipes, and so it is

107 Recently shredded oak, Varese Ligure, 2002.

possible that the species was fairly common during the period of wood pasture. In areas where the growth of the incoming hop hornbeam was producing greater shade, species such as juniper were becoming suppressed. The recent history of the farm shows how vegetation, following a long period of intense management at least from the mid-eighteenth century through to the mid-nineteenth, responded to a period of almost complete abandonment in the last third of the twentieth century and emphasizes how rapidly the managed landscape can be replaced by dense shrubs and trees.

Terrace abandonment and tree growth

The Lagorara Valley is a small valley near Maissana in the Val di Vara. Much of the valley has step terraces with stone walls and the cadastral map indicates that the ownership is extremely fragmented, with hundreds of small plots of less than 500 sq metres (5,382 sq ft). An old mule track runs up the valley and there are many small wooden and stone barns, a few of which even today are used for sheep. But the mule track, like most in the Val di Vara, is now largely overgrown and access to the valley is provided by a track for wheeled vehicles built in the 1970s. A local farmer explained how until the 1960s the terraces were used to grow potatoes, maize, wheat and barley, but after that time this became uneconomic. By the late 1990s many of the plots were abandoned but most were freely grazed by sheep between April and September. Some terraces were burned every two years to improve the quality of the grazing; the burning stops the spread of coarse grasses such as *Brachypodium rupestre* and shrubs such the tree heather. The slopes became steeper close to the river and these largely abandoned areas had many naturally regenerating trees such as hop hornbeam and manna ash.

The survival of a photograph taken in about 1955 allows us to see very distinctly some of the terraces around a small group of buildings at Casoni on the slopes of M. Porcile in the Lagorara Valley. The photograph shows well-maintained buildings and clearly defined terraces that are neatly kept. Many of the trees have been shredded and a few only have spindly branches growing at their tops. A local resident stated that the terraces had been cultivated for wheat, maize and potatoes at this period. The last 50 years have seen the complete abandonment of this part of the valley. The buildings themselves have almost disappeared and most of the walls have collapsed. Almost all the terraces so clearly depicted in the photograph have been abandoned for at least 30 years and are now overgrown by dense vegetation. Species such as blackthorn, tree heather and hop hornbeam have rapidly encroached, and trees now dominate the landscape.

We studied two sets of terraces in the parish of Valletti on the other side of M. Porcile in 2002 at Bonello and La Vignola to see how rapid the spread of trees onto cultivated terraces could be. These terraces probably date from at least the eighteenth century and their extent is shown relatively clearly on a postcard dating from *c.* 1965 where Valletti is photographed from the other side of the Val di Vara. The flight of terraces follows the spur on the north-facing slopes of the valley. The two sets of terraces are set in a matrix of old chestnut woodland charac-terized by pocket terraces; some of the chestnut trees growing in the stone walls are probably 300 years old.

The abandonment of the terraces can be documented through oral history and the use of field survey. Maria has lived in Valletti since 1931 and she and her husband used the terraces at Bonello In Fundo a e Tere for growing potatoes, maize, fruit trees, beans and winter cabbage. The terraces were also used to grow hay, but cultivation ceased and the terraces were abandoned around 1992. This oral history was backed up by the field evidence. Many of the terraces were covered by dense areas of *Brachypodium rupestre* with other species including *Festuca rubra*, *Dactylis glomerata* and St John's wort. The shrub and tree species varied depending on the length of abandonment and the location of seed trees.

108 Lagorara valley, near Maissana, Liguria, in a photograph of *c.* 1955.

109 Willow pollards at Valletti, Liguria, 1997.

The main shrub species were roses, old man's beard and brambles. The higher terraces had many seedlings of sycamore derived from a tree in the village. The lower terraces were dominated by extensive, dense patches of suckers from clones of cherry and aspen which were dated at between twelve and fourteen years old by counting the tree rings. Further below these terraces was woodland with many old chestnut stools and a herb layer including wood anemone and bilberry.

At Vignola Davide told us that his terraces were cultivated and used to grow grain, maize and potatoes on a rotational basis until the 1960s. Since then the terraces had been used to grow hay. Some fruit trees and vines remained and there are characteristic small willow pollards that are still cropped annually to produce willow ties for the vines. Davide left the village in 1964, but returned when he retired and tries to 'keep his terraces tidy'. The higher terraces near the house are cut for hay and the species include sorrel, buttercup, white clover and ground elder. The middle terraces were unmanaged and had tall overgrown grass with a considerable build-up of dead grass and stems. Some were dominated with rose species, blackthorn and brambles, but most were dominated by very dense *Brachypodium rupestre*. There was considerable growth of suckers from domestic plum and cherry trees. The lowest terraces had been abandoned for up to 40 years; they still appear open on the post-card view of *c.* 1965 but in 2002 consisted of fairly dense woodland made up of young chestnut, hazel and hop hornbeam. The herb layer had wood anemone and ivy. These two examples at Valletti demonstrate the remarkable speed with which trees can colonize abandoned formerly cultivated terraces. The first stage is often for suckers from cultivated fruit trees, such as plum and cherry, that are frequently grown on the edges of terraces, which spread rapidly through the untilled soil. The ground vegetation very soon becomes dominated by a thick and coarse sward of *Brachypodium*. This is soon overcome by dense thickets of rose and blackthorn and by seedlings from trees such as the hop hornbeam. Within fifteen years dense, impenetrable woodland has appeared, camouflaging the remnants of terrace walls, paths and sheds, which are all that remains to show the former intensive agricultural activity.

Abandoned chestnut woodland

If one had to choose one tree that symbolized Liguria it would have to be the chestnut. Not only are vast areas of the mountainsides covered with chestnut woodland, but there are many ancient chestnut trees and everyone knows that the cultivation of the chestnut was once so important that it was the staple crop in the hills, with chestnut flour, bread and pasta being of crucial importance in sustaining the population up until the middle years of the twentieth century. The importance of chestnuts in the cultural and social history of Liguria and the many notable ancient trees suggest that chestnuts have been present from time immemorial, and research suggests that this may indeed be the case. A recent analysis of pollen evidence concluded that there were six main regions that acted as glacial refuges where the tree could survive the last ice ages. One of these was in 'central and southern Italy extending along a constricted hilly belt between the Tyrrhenian coast and the Apennine ridge with a possible extension towards the north', including the 'Ligurian Apennine, Cuneo region'.[8] However, a detailed investigation of pollen evidence at four sites in the upper Trebbia and Aveto valleys (between 812 metres and 1,331 metres) suggests that chestnuts were not present until the Roman period, 'when the general reduction in woodland cover led to the formation of a new and distinctive vegetation community dominated by non-arboreal pollen' and trees such as olives, walnuts, chestnuts and pines, indicating cultivation.[9] Whether classed as native or not, the evidence suggests that 'the greatest interest in the management of chestnut for fruit production developed after the Roman period and can be associated with the socio-economic structures of medieval times.'[10]

Monastic land charters of the eleventh century and archaeological reports indicate that chestnuts were of great importance in the
Middle Ages. In charters they were described as *castanea*, meaning a tree,
and sometimes a named individual tree, or as *castanetum*, a chestnut
woodland or plantation. A contract drawn up in November 1006 between
Martin, who lived at Gallaneto in the Polcevera Valley, and the monastery
of Santo Stefano in Genoa, specified that he had to 'prepare the ground
for chestnut trees and improve [them] and to put domestic chestnut
trees where appropriate' so that after ten years had passed he could pass
half the crop to the monastery and keep the remainder himself. This
shows how areas of chestnuts might be 'improved', probably through
grafting, to increase the quality and quantity of the crop. The central
place of chestnuts to medieval communities is indicated by recent excavations in the lower Vara valley at the old hillside village of Corvara.
Here the 'finds in the eleventh-century phase all pointed to a mixed
economy, largely self-sufficient, based on the chestnut'. These included
two grinding stones for chestnut flour and many fragments of pottery
testi, a characteristic, locally made small baking pot of a type used to
make chestnut-flour bread until at least the 1950s.[11]

The chestnut groves and orchards were celebrated by visitors and
travellers to Liguria in the nineteenth century, none more so than Alice
Comyns Carr, who in her travel book *North Italian Folk: Sketches of Town
and Country Life* (1878) provided rich descriptions of the chestnut harvest. She gives a full and realistic account of the process of harvesting
chestnuts, the tools used and the sounds of the nuts falling to ground
and of the harvesters talking and singing. This section is illustrated by
a drawing made by Randolph Caldecott of women gathering chestnuts with the use of pincers. Caldecott (1846–1886) was an enormously
successful artist and book illustrator who specialized in rural scenes and
activities and travelled extensively in Italy. He carefully depicts the floor
of the chestnut grove as characteristically kept clear of shrubby vegetation by grazing, supplemented by careful sweeping and brushing,
which assisted with gathering the nuts.

From the fifteenth to the nineteenth century chestnuts were also
pollarded to allow the production of charcoal for the Genoese iron
industries. This could be combined with nut production by the selec-
tive felling of those branches which had reached about 7 centimetres
(2¾ inches) in diameter, known as *rami da carbone*, leaving behind the
branches that produced fruit. Some workers thought that this practice

encouraged the 'precocious production of fruit crops', but it died out in the late nineteenth century.[12] Chestnut flour began to become unfashionable in the early twentieth century with the spread of wheat flour and commercially produced dried pasta. This coincided with a rapid rate of rural depopulation and crucially, for the chestnut, the arrival from America of *Cryphonectria parasitica* (formerly known as *Endothia parasitica*), the chestnut blight. The disease first appeared in Europe in 1938 at Genoa, and it spread rapidly through the chestnut-growing parts of Italy and soon reached Spain and France. Now it is common throughout Europe, the only exceptions being the Netherlands and Britain. Unlike in America, where the blight kills the trees, in Italy and Europe more widely individual trees have recovered from the disease due to the 'natural occurrence of hypovirulence' indicated by the presence of healing cankers and scars, which are clearly visible on many chestnut trees throughout Liguria.[13]

Many chestnuts were grafted so as to improve the quality of the fruit produced, and they were carefully grown on terraces even on very steep slopes. In a few places in the Val di Vara, examples of traditional,

110 Randolph Caldecott, 'Gathering the Chestnuts', from Mrs Comyns Carr, *North Italian Folk: Sketches of Town and Country Life* (1878).

carefully pruned trees associated with small-scale cultivation can still be seen. In most of the valley the combination of chestnut blight and the collapse of the market for chestnuts and flour has led to the abandonment of many chestnut orchards. Some have been converted into commercially viable areas of dense chestnut coppice, but large areas of neglected chestnut groves survive that contain important evidence of the past modes of culture which had been important for hundreds of years. The chestnuts needed to be dried, and often small buildings called *alberghi* or *casun* were built for this purpose. These had a drying floor made of slats of alder wood on which the nuts were piled. Below this a fire would be lit, and the hot air would filter out of small holes left in the wall above the chestnut loft. These small buildings can be found as isolated structures near to chestnut woods, but were also constructed adjoining or as part of rural houses. They can usually be easily identified today by the wooden slats.

We investigated in March 2003 one site at Palarino, Teviggio, which had a derelict chestnut-drying building or *casun* and nearby some completely abandoned chestnut woodland where the local geology is mainly Monte Zatta sandstone. Although marked on modern maps as Palarino, local people do not call it by this name but refer to the unenclosed parts simply as *communeglia*, or common land. On the vegetation map the area is on the border of two zones, one characterized by *vegetazione arbustiva*, with species such as *Calluna vulgaris*, *Erica carnea* and *Genista salzmannii*; the other *vegetazione arborea*, characterized by chestnut and Turkey oak.[14] Neighbouring farmers told us about the local history of the site and about the disused water channel that had previously been used to transport water to a neighbouring farm. There were several such water channels in the district, called *surku* in local dialect. The *surku* stopped being used in the 1970s when the open channel was replaced by a black plastic water pipe.

The chestnut woodland was on acid sandy soil and was mainly composed of abandoned chestnut coppice. The form of the coppice stools indicated that the original main stem had been cut many years ago. The age of these original standard trees was estimated from the size of the stumps to be over 100 years. This ties in with the historical map evidence, which indicates that the site was wooded in both 1853 and 1936. The most recent small-scale coppicing had taken place about twelve years before 2003. In general, however, the younger stems growing around the old main stem were estimated to be around 30–40 years old. Many of the chestnut stems were clearly infected with chestnut blight. The other tree species found at the site included manna ash, juniper and Turkey oak. These were mainly young saplings of between twelve and nineteen years of age. The shrub species included *Calluna vulgaris*, *Erica arborea*, *Erica carnea* and *Cytisus scoparius*. Several of these shrubs had reached a height of about 2 metres (6½ feet). The ground layer was sparse, with some grasses such as *Brachypodium pinnatum* and *Thymus* spp.

The *casun* on the edge of the chestnut area was built at the bottom of a small series of south-facing abandoned terraces.[15] The landowner indicated that the terraces had been used for the cultivation of potatoes and other crops until the 1960s; after that they have been used for grazing. The area was fenced and sheep grazed the area until recently; most of the young trees growing on the abandoned terraces were three years old. Searches for pottery fragments on the terraces, especially on the edges of terraces that were eroded, provided several shards of taches-noir decorated terracotta, which was produced at Albisola near Genoa. At the end of the eighteenth century it was characteristically a brown-orange colour while in the early nineteenth century black pottery became more common. The shards will have derived from broken pottery carried to the terraces with domestic waste used as manure. Shards of both orange-brown and black pottery were found, indicating that the terraces were likely to have been constructed by at least the turn of the eighteenth century. The *casun* was shown on both the 1978 and 1936 maps. Inside there was evidence of a former loft and a ladder. The stonework of the building was similar in style to that of other buildings, such as the farm at Cunie, which had been built in the late eighteenth century. Although the *casun* had recently been used for storing hay and animals, the evidence of the slatted wood in the loft and the burn marks on the old beams indicated that it had also been used for drying chestnuts.

This chestnut stand was used for the production of chestnuts until the end of the nineteenth century. The chestnuts are likely to have been dried in the eighteenth-century *casun*. With the rapid decline in population in the twentieth century and the arrival of chestnut blight, the economic production of chestnuts was not possible. Most of the standard chestnut trees were felled, especially during and immediately after the Second World War, and coppice regrowth developed. Coppicing continued until the 1960s. There may have been sporadic grazing of the area, but this stopped about twenty years ago, allowing natural regeneration of shrubs and trees. The abandonment of the *surku* in the 1970s and the bulldozing of a new road adjoining the site reduced the water supply to the site. The terraces had been constructed by the late eighteenth century and were cropped until the 1960s. Individual trees on the terrace were cultivated for chestnuts and oak leaf fodder. After a period of about 40 years when the terraces were used for grazing, this use has now ceased, and natural regeneration of shrubs and trees is occurring. Since 2003 the complete abandonment has continued: the *casun* largely collapsed in 2007 and the dense natural regeneration of shrubs and trees increasingly conceals the evidence of chestnut culture.

This type of abandoned and derelict chestnut landscape is highly characteristic of Liguria, but the specialist growing of chestnuts does survive in some parts. In the upper Bormida valley in Savona, for example, commercial chestnut orchards were established in the medieval period and some farmers still grow the local variety of chestnut known as the *gabbiana*. This variety was chosen by the Slow Food Association as one of its quality products in 2002. The chestnuts are dried in stone or brick buildings known locally as *tecci*, which can stand alone or form part of a house. They consist of a single room with a central open chimney and a loft made of wooden slats called *graia* about 2–3 metres (6½–9¼ feet) from the ground. A cooperative group of local growers called Il Teccio are promoting the ancient techniques of growing and gathering chestnuts and have obtained Protected Geographical Indicator status for *Castagna essiccata nei tecci di Calizzano e Murialdo*. But some of the traditional practices, such as the grazing of sheep within the orchards to keep the ground clear, are no longer practised, which 'not only diminishes the beauty of the orchards and makes chestnut gathering more difficult' but is seen as increasing 'the risks of parasites and diseases' and allowing the conversion of orchards into woods. In the upper Sturla Valley at Borzonasca there is a good example of well-maintained terraced chestnut groves. These are irrigated by a complicated system of aqueducts, which are documented from at least the end of the seventeenth century. Some local growers have set up a cooperative, Il Castagno, which promotes the commercial growth of the chestnut of Borzonasca, but even here the traditional management of terraces is threatened by an ageing population and by the abandonment of the chestnut groves, leading to their conversion into mixed woods and the loss of their historical and cultural characteristics.[16]

Rural depopulation and land abandonment can lead remarkably rapidly to dramatic changes in the woodland cover of a region. In many parts of Liguria an unintentional consequence of abandonment has been the natural regeneration of dense areas of woodland which forms an example of unplanned rewilding. A small area of woodland can contain trees with very different management histories: old trees that are marked by ancient management practices such as shredding, which have been carried out for many hundreds of years, are frequently surrounded by burgeoning regrowth of young trees and shrubs. This unplanned semi-natural vegetation blankets the landscape rather more effectively than the plantations of trees that were made in several parts of the area during the twentieth century.

The rapid spread of new semi-natural woodland is not restricted to Liguria. The same is happening in many of the more mountainous parts of Italy and other Mediterranean countries and is beginning to happen in those East European countries where agriculture is currently being modernized. This unplanned and unprecedented rapid woodland expansion is seen by some as a welcome return to some sort of idealized past, before the agricultural and pastoral activities of humans started to manage and mould trees and woods thousands of years ago. The idea of rewilding former agricultural and pastoral areas is becoming increasingly popular. In Britain the three bodies which own Ennerdale in the Lake District (the Forestry Commission, the National Trust and United Utilities) have set up the Wild Ennerdale Project and a management plan was established in 2006 'to allow the evolution of Ennerdale as a wild valley for the benefit of people'. The idea is not to recreate 'some past state' but to use natural processes to allow changes in the 4,300 ha (10,625 acres) of upland grazing, broadleaved and coniferous woodland. In 2011 Rewilding Europe, a Dutch foundation, was established; it aims to rewild a million ha (2,471,000 acres) in Europe by 2020.[17]

In Liguria the consequences of decades of unplanned and unmanaged rewilding are readily apparent. The new woodland that now clothes the hills has allowed the successful spread of large numbers of wild boar and smaller numbers of deer and wolves, which had been erased from the landscape by intense cultivation, trapping and shooting in the nineteenth century. This new natural woodland is therefore thought by some conservationists to be a major success story, allowing the re-introduction of rare mammals and birds and the production of some valuable timber and possibly even making a small contribution towards carbon sequestration. But others recognize that the new woodland disguises the loss of traditional woodland knowledge, the disappearance of thousands of small agricultural terraces and the reduction of biodiversity associated with the pastures and meadows of the cultural landscape it replaces.

Afterword

THE example of Ligurian woodlands shows how, through inaction, humans have allowed the silent and stealthy spread of trees over thousands of acres of the Italian Apennines. The same process is going on in many parts of the developed world where the pressure of grazing and cultivation is reduced. Vast tracts of Europe and America which were once agriculturally productive are now covered in semi-natural woodland. This localized relaxation of agricultural pressure is taking place in a broader context of agricultural intensification. Rural landscapes are bifurcating into areas dominated by modern agriculture producing heavy yields of rice, barley and wheat, such as the Po Valley of northern Italy and the arable landscapes of eastern England, and the largely abandoned mountain slopes of the Apennines and the Alps, which are rapidly being covered with trees. On a smaller scale many areas of English common land which were grazed by geese, sheep, cattle and horses until the mid-twentieth century soon became overspread with shrubs and trees within a few years of the grazing ceasing and are now dense woodland. Formerly grazed Surrey heathlands have become dominated by birches and pines. Abandoned riverside meadows once cut annually for hay are very soon overgrown by willows and alders. Wherever grazing, cultivation or the cutting of hay cease, trees colonize the land and vigorously dominate the landscape.[1]

Trees have life histories of their own and within woods and forests they coexist with other species and influence each other. But the dominance of human activity means that their form and extent are largely contained by human activities. As soon as trees become a threat, for example when they grow near houses or transport routes, they are controlled through pruning and pollarding. One of the most common reasons for coppicing trees today is to stop trees overgrowing power lines and causing electricity cuts during storms. The paths of power lines can easily be traced in wooded mountain scenery by the narrow belts of coppiced woodland that lies under the pylon and cables. The relationship between trees and people is thousands of years old and human knowledge of trees is extensive. Trees and forests as oxygen producers enable life on earth as we know it to continue. They were crucial to the development of human civilization. Our knowledge of techniques of reproducing, cultivating and growing different species of trees in different places is now vast, as is the ability to select, breed, hybridize and genetically manipulate trees for particular characteristics. But as soon as trees are released into the landscape the consequences are uncertain.[2]

We can take *Robinia pseudoacacia,* known as the false acacia or black locust tree, as an example. This tree is native to the eastern USA and has been widely planted around the world. It was introduced into Britain in the seventeenth century and soon attracted considerable attention. Its cause was taken up with tremendous enthusiasm by William Cobbett (1763–1835), the writer and farmer from Farnham most famous as editor of the *Political Register* and author of *Rural Rides.* It was for him 'the tree of trees' and should be planted by everyone who had land. But a contemporary marginal note written in pencil in my first edition of Cobbett's *The Woodlands* (1825) shows how this optimism was misplaced: 'N. B. Notwithstanding all that has been said by Cobbett in favour of the Locust, it is certain that the wood of this tree is next to *worthless.* His account of it is much exaggerated. In England it is only suitable for ornament in sheltered places.' Indeed, it was never planted widely as a woodland tree in Britain but is now, especially in the golden-leaved form 'Frisia', a very popular garden tree. It has several very bad characteristics, one of which is unexpectedly dropping its brittle branches. Another is its very spiny branches which make cutting and felling dangerous. The main problem, however, is its very vigorous suckering habitat, which is difficult to keep in check. The tree has been widely planted in Europe, South Africa and Japan, and has been particularly favoured as a tree to help consolidate road and railway embankments and cuttings. But it vigorous vegetative spread means that it is now identified as a serious and invasive pest tree which is very costly if not impossible to eradicate (illus. 77).[3]

The demands we place on trees change over time and the potential benefits of encouraging a particular species foreseen by one generation are commonly no longer relevant or apparent when the trees have become established. New values are assigned to trees. Ancient trees are increasingly being valued as repositories of environmental knowledge and tree rings are examined to gain knowledge of fluctuations in temperature and likely regional histories of climate change. Dendrochronological studies also allow the peopling of past landscapes. At Loch Katrine, Forestry Commission Scotland has cored ancient ash trees in the wood pastures on the southeastern shores of the lake. No one now lives in the area, which was once so important for cattle and sheep farming and whose importance for hunting was captured in Walter Scott's poem 'The Lady of the Lake' (1810). The detailed analysis of the cores for two ash trees showed that both were of late seventeenth-century origin and that one had been regularly pollarded in the eighteenth century. Several of the alder and oak trees cored revealed that trees of nineteenth-century origin had originally been coppiced, but the coppice regrowth had fused together to form what appeared to be a single stem. This dendrochronological work is uncovering unknown tree and woodland histories.[4] Genetic analysis of trees is one of the areas that has most potential for our understanding of long-term woodland history. One of the most striking examples is the genetic history of the English elm (*Ulmus procera*) and the wych elm (*U. glabra*). Analysis of samples taken from elms in Spain, France, Greece, Italy and Britain established the variability of chloroplast DNA and likely genetic lineages. A clone of *U. procera* common to Italy, Spain and Britain has been identified. This elm only rarely produces seed, but spreads by suckers very well and was once one of the commonest hedgerow trees in England. It had long been thought that the tree had been introduced by the Romans or even earlier during the Bronze Age, but the genetic research suggests that the clone is the same as the Atinian elm from Latium recommended by Columella as vine supports in *De Re Rusticus* around AD 50.[5]

Woods do not have to be exciting or contain rare species to be interesting and all have distinct histories and unknowable futures. Dukes Wood, found on the clay soils of central Nottinghamshire, is a small and insignificant ancient wood managed as a nature reserve by Nottinghamshire Wildlife Trust. It is only 8 hectares (19 acres) in area and many people would deem it nondescript. The wood is on the borders of the parishes of Eakring and Winkburn and ash, elm, oak, hazel and birch are the main trees, with a rich ground flora of bluebells, primroses, wood anemone and yellow archangel. It has like most of the surrounding small woods been managed for centuries largely under the system known

111 Overgrown
ash coppice,
Dukes Wood,
Nottinghamshire,
2013.

as coppice with standards. These coppice woods were known in the eighteenth and nineteenth centuries as spring woods, and George Sanderson's *Map of the Country Twenty Miles Round Mansfield* of 1835, shows names such as 'Redgate Spring' and 'Orchard Wood and Nut Wood Spring.' In the 1790s the Winkburn Woods were coppiced every twenty years and were being replanted with ash trees for this purpose.[6] Many of the poles cut were used as hop poles and this was a lucrative market as hop growing was one of the leading agricultural activities in east Nottinghamshire in that period. Dukes Wood is too small to be named on the map but is shown just to the west of Dilliner Wood. The continuity implied by the old map and the current woodland disguises a remarkable episode of its history. Between 1939 and 1966 Dukes Wood and the area around it was an important onshore oilfield. The trees helped to camouflage, perhaps more theoretically than in practice, the oil workings from wartime German aerial photography. Almost all the roadways, storage tanks and concrete surrounding the wells have now been removed but a few of the pumps known as nodding donkeys have been restored and there is a small Dukes Wood Oil Museum. But these are the only reminders of the localized oil boom; a walk through the wood today provides hardly any clues to this 30-year period of intense industrial exploitation.

What do the trees of Dukes Wood tell us of its history and future? Many of the older trees were once coppiced and this large ash tree has six main stems and was probably last coppiced in the 1960s. In Robert Monteath's book *The Forester's Guide and Profitable Planter* of 1824 there is an engraving which shows the growth of ash coppice at 15, 20, 25 and 30 years following coppicing. Coppicing also produced firewood and many useful wooden products, but as we have seen by the beginning of the twentieth century the markets for most of these products had disappeared. Many woods became uneconomic and tended to be more important for preserving game than producing income from coppicing. Once abandoned the growth of large stems from old coppice stools often makes them unstable, and eventually some or all of the stems are likely to collapse. One way to encourage the longevity of such stools is to re-coppice them, and this has become a standard nature conservation practice. Moreover the market for ash firewood has recently become more buoyant than for many years and there is some hope that the popularity of wood-burning stoves will restore dynamic management to such woods and help pay for conservation management.

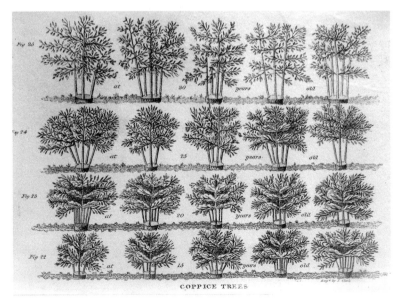

112 'Coppice Trees at 15 . . . 20 . . . 25 [and] 30 years old', engraving from Robert Monteath, *The Forester's Guide and Profitable Planter* (1824).

There are enormous numbers of young ash trees naturally regenerating throughout Dukes Wood. Ash trees are prolific seed bearers and
these germinate and grow easily in the clay soils. The ability of ash trees
to regenerate freely, the increase in the demand for firewood and the
potential for growing high-quality ash timber meant that the future for
ash silviculture was very optimistic until very recently. However, with
the identification of ash die back disease (*Chalara fraxinea*) for the first
time in the UK in the summer of 2012 the future of the tree is a matter
of great uncertainty. The disease was first noticed in Poland in 1992 and
it spread rapidly. The Forestry Commission initially thought that it had
arrived in Britain on young trees 'imported from nurseries in Continental
Europe' but the infection was soon found in older trees in East Anglia
and Kent and it may have been 'carried on the wind or on birds coming
across the North Sea and English Channel' or by people or machinery
who had visited infected Continental sites. No one knows how the
spread of the disease will develop over the next few years but, taking into
account what has happened in other countries, it is likely that the
spread of *C. fraxinea* will be rapid and it will affect many trees.[7]

Elm trees are frequent in Dukes Wood today; most are fairly young,
around 20–30 years of age, and many appear in rude health. But these
trees are also subject to a virulent disease. Those growing today have
sprung up since the arrival of Dutch Elm Disease in Nottinghamshire
in the early 1970s. By the late 1970s almost all elm trees, whether in
hedgerows or woods, or whether the English elm (*Ulmus procera*) or the
Wych elm (*U. glabra*), were dead or dying. But since that time many
elm trees have regrown. The disease has by no means disappeared
however: it tends to reappear once the new generation of elm trees has
become large enough to support the breeding of the beetles which
carry the disease. This is what is now happening at Dukes Wood. A fair
number of the new generation of elm trees are dead or dying and there
is clear evidence of the tunnels bored by elm beetles in bark which has
fallen away from the trees. The leading plant pathologist Clive Brasier
has pointed out that 'woodlands and landscapes in the UK and across
the world are suffering from pathogens introduced by human activities',
and there is little that can be done to stop this spread. The death of most
ash trees in Dukes Wood, as with the elms 40 years ago, would bring
about significant changes in the short term, but in the longer term it
would provide space for other species such as alder, birch, field maple
and oak to regenerate.

The particular history and uncertain future of Dukes Wood display on a small scale the complex relationship between people, trees and woods. The diverse episodes and themes of this book show how we have always relied on trees for sustenance and pleasure and how the fate of many trees and woods is very much dependent on human activities. People select certain trees for particular purposes and markets and shape the form of individual trees through management, but they also help spread virulent tree diseases. To understand individual woods and trees it is necessary to have knowledge of their precise history and geography. But a grasp of the implications of global movements and processes is also essential. Trees, woods and forests appear on the surface to be stable and unchanging features against which we can match our individual lives and the lives of nations and civilizations. But the history of trees is constantly being rewritten and the future of trees is uncertain.

References

Introduction

1 P. Jones, 'The Geography of Dutch Elm Disease in Britain', *Transactions of the Institute of British Geographers New Series*, 6 (1981), pp. 324–36; Isobel Tomlinson and Clive Potter, '"Too Little, Too Late"? Science, Policy and Dutch Elm Disease in the UK', *Journal of Historical Geography*, 36 (2010), pp. 121–31.
2 D. A. Stroud et al., *Birds, Bogs and Forestry* (Peterborough, 1987).
3 P. Schutt and E. P. Cowling, 'Waldsterben, a General Decline of Forests in Central Europe: Symptoms, Development, and Possible Causes', *Plant Disease*, 69 (1985), pp. 548–58; J. M. Skelly and J. L. Innes, 'Waldsterben in the Forests of Central Europe and Eastern North America: Fantasy or Reality?', *Plant Disease*, 78 (1994), pp. 1021–32.
4 Oliver Rackham, *Trees and Woodlands in the British Landscape* (London, 1976); Rackham, *Ancient Woodland* (London, 1980); George Peterken, *Woodland Conservation and Management* (London, 1981); Charles Watkins, *Woodland Management and Conservation* (Newton Abbot, 1990).
5 Robert Mendick, 'How England's Forests Were Saved for the Nation', *Daily Telegraph* (19 February 2011); Forestry Commission, *Independent Panel on Forestry: Final Report* (London, 2012).

ONE: Ancient Practices

1 Konrad Spindler, *The Man in the Ice* (London, 1994); Klaus Oeggl, 'The Significance of the Tyrolean Iceman for the Archaeobotany of Central Europe', *Vegetation History and Archaeobotany*, 18 (2009), pp. 1–11.
2 Oeggl, 'The Significance of the Tyrolean Iceman', p. 3.
3 Spindler, *Man in the Ice*, pp. 87–8, 218.
4 Oeggl, 'The Significance of the Tyrolean Iceman', p. 3.
5 Ibid., p. 5.
6 Bryony Coles and John Coles, *Sweet Track to Glastonbury: The Somerset Levels in Prehistory* (London, 1986); C. W. Dymond, 'The Abbot's Way', *Somerset Archaeology and Natural History*, 26 (1880), pp. 107–16.
7 Coles, *Sweet Track*, p. 31.

8 Ibid., p. 44.

9 Ibid., pp. 45, 50.

10 Ibid., pp. 51, 55.

11 Maisie Taylor, 'A Summary of Previous Work on Wood and Wood-Working', in *Flag Fen: Peterborough Excavation and Research, 1995–2007*, ed. Francis Pryor and Michael Bamforth (Oxford, 2010), p. 69; Michael Bamforth, 'Aspects of Wood, Timber and Woodworking at Flag Fen, 1995–2005', in *Flag Fen*, ed. Pryor and Bamforth, p. 72; Francis Pryor, *Flag Fen* (Stroud, 2005), p. 132.

12 Maisie Taylor, 'Big Trees and Monumental Timbers', in *Flag Fen*, ed. Pryor and Bamforth, p. 90; Russell Meiggs, *Trees and Timber in the Ancient Mediterranean World* (Oxford, 1982), pp. 346–7; Pliny the Elder, *Natural History*, vols XII–XVI, trans. H. Rackham (Harvard, 1956), vol. XVI, pp. 83, 227.

13 Taylor, 'Big Trees', p. 92.

14 Ibid., p. 94.

15 Pliny the Younger, 'Letter 6 to Domitius Apollinaris', in *Complete Letters*, trans. P. G. Walsh (Oxford, 2006), Book V.

16 Meiggs, *Trees*, pp. 269–70 and translations of Pliny the Younger, V, 6; III; III, 19.

17 Pliny the Elder, *Natural History*, XIV, 1, p. 187.

18 Ibid., XII, 1, pp. 3–5.

19 Ibid., pp. 8–9.

20 Ibid., XV, 20, p. 341.

21 Ibid., XVI, 2, pp. 389–91.

22 Meiggs, *Trees*, p. 270.

23 Ibid., p. 262.

24 Ibid., pp. 265–6.

25 A jugerum = 0.25 hectare/⅝ acre.

26 Meiggs, *Trees*, pp. 18–19.

27 Ibid., pp. 118–19.

28 Edward Forster, 'Trees and Plants in Homer', *Classical Review*, 50 (1936), pp. 97–104; Meiggs, *Trees*, pp. 106–10.

29 *Iliad*, trans. A. T. Murray and revd William F. Wyatt (London and Cambridge, MA, 1999), XIII, 389; the same simile is used when Patroclus kills Sarpedon, ibid., XVI, 482.

30 Ibid., IV, 480.

31 Ibid., XIII, 178.

32 Ibid., XVI, 113, 141.

33 Ibid., XIV, 395.

34 Ibid., XVI, 767.

35 Ibid., XII, 31.

36 Ibid., XI, 86.

37 Ibid., XXIII, 315.

38 *Odyssey*, trans. A. T. Murray and George E. Dimock (London and Cambridge, MA, 1919), X, 210, 241–4.

39 *Iliad*, XII, 141.

40 *Odyssey*, V, 11, 106.

41 *Iliad*, II, 306.

42 Herodotus, trans. A. D. Godley (London and Cambridge, MA, 1922), VII, p. 31.

43 Pausanias, *Description of Greece*, trans. W.H.S. Jones, (London and Cambridge, MA, 1933), XI, Elis II, XXIII, 1; Meiggs, *Trees*, p. 273; *Pausanias's Description of Greece*, trans. J. G. Frazier (London, 1898), III, XIV, 8.

44 Pliny the Younger, 'Letter 6 to Domitius Apollinaris' (Oxford, 2006), Book V.

45 Pliny, *Natural History*, XIXIX, pp. 91–102.

46 Ibid., and Strabo, *Geography*, trans. Horace Leonard Jones (London and Cambridge, MA, 1923), Books 4, 6, 2; 17, 3, 33–4; Lucan, *The Civil War*, trans. J. D. Duff (London and Cambridge, MA, 1928) IX, 426–30.

two: Forests and Spectacle

1 Russell Meiggs, *Trees and Timber in the Ancient Mediterranean World* (Oxford, 1982), p. 82.

2 Manolis Andronicos, *Vergina: The Royal Tombs and Ancient City* (Athens, 1984), pp. 106–19; Konstantinos Faridis, *Vergina* (Thessaloniki, n.d.), pp. 37–8; James Davidson, 'Crowning Controversies', *Times Literary Supplement* (20 May 2011), pp. 17–18.

3 Quintus Curtius Rufus, *History of Alexander*, trans. John C. Rolf (London and Cambridge, MA, 1946), 8, 1, 11–16; Meiggs, *Trees*, p. 272.

4 Anthony Birley, *Hadrian: The Restless Emperor* (London, 1997) pp. 25, 137, 164; Thorsten Opper, *Hadrian: Empire and Conflict* (London, 2008) p. 242, fn 19.

5 Birley, *Hadrian,* pp. 240–41.

6 Opper, *Hadrian*, pp. 171–2; BM GR 1805.0703.121.

7 See Society of Antiquaries of London Online Newsletter (SALON), www.sal.org.uk/salon.

8 Sir James Frazer, *The Golden Bough: A Study in Magic and Religion* (London, 1922), pp. 1–2, 701.

9 Corinne Saunders, *The Forest of Medieval Romance* (Woodbridge, 1993), pp. 24, 21, quoting Servius' commentary of the *Aeneid*.

10 Virgil, *Aeneid*, trans. H. Rushton Fairclough (London and Cambridge, MA, 1935), VI, 175–90, 205–11.

11 Ibid., I, 162–6.

12 Ibid., IX, 375–92.

13 Saunders, *Forest,* p. 26.

14 Virgil, *Eclogues*, I, 1–25.

15 Ibid., VII, 53–67.

16 'Terrestrial Paradise', *Catholic Encyclopaedia*.

17 R. I. Page, *Norse Myths* (London, 1990), p. 59.

18 Jonas Kristjansson, *Eddas and Sagas* (Reykjavik, 1997), p. 41.

19 E.O.G. Turville-Petre, *Myth and Religion of the North* (London, 1964), p. 277.

20 Page, *Norse Myths*, p. 59.

21 Turville-Petre, *Myth and Religion,* pp. 42–3.

22 Ibid., pp. 48–50.
23 Ibid., pp. 182–3, 244–6.
24 Saunders, *Forest,* pp. 10–11.
25 Ibid., pp. 11, 17–18.
26 James Holt, 'The Origins of Robin Hood', *Past and Present* (November 1958), p. 37.
27 Chrétien de Troyes, *Arthurian Romances,* trans. W. W. Comfort (London, 1975), pp. 216–17.
28 Saunders, *Forest,* p. 10.
29 Della Hooke, *Trees in Anglo-Saxon England: Literature, Lore and Landscape* (Woodbridge, 2010), p. 142.
30 David Bates, 'William I (1027/8–1087)', *Oxford Dictionary of National Biography* (Oxford, 2004).
31 Frank Barlow, 'William II (c. 1060–1100)', *Oxford Dictionary of National Biography.*
32 Frank Barlow, 'Henry I (1068/9–1135)', *Oxford Dictionary of National Biography.*
33 Charles R. Young, *The Royal Forests of Medieval England* (Pennsylvania, 1979), pp. 17, 19.
34 Thomas K. Keefe, 'Henry II (1133–1189)', *Oxford Dictionary of National Biography.*
35 Young, *Royal Forests*, pp. 68–9.
36 Jean Birrell, 'Records of Feckenham Forest, Worcestershire, c. 1236–1377', *Worcestershire Historical Society New Series,* 21 (2006), p. xiv.
37 Birrell, 'Records', p. xiv.
38 Ibid., p. 159.
39 Oliver Rackham, *Ancient Woodland* (London, 1980), p. 177.
40 Stephen A. Mileson, *Parks in Medieval England* (Oxford, 2009), p. 30.
41 Robert Hearn, 'Grey Wolves and Wild Boar: Comparative Species History in Liguria, 1500 to 2012', PhD thesis (University of Nottingham, 2013).
42 Birrell, 'Records', p. xiv; Roger Lovegrove, *Silent Fields* (London, 2007), p. 22; Birrell, 'Records', p. 67; Oliver Rackham, *History of the Countryside* (London, 1986), p. 36.
43 Birrell, 'Records', p. xv.
44 Ibid., p. xvii.
45 Ibid., p. xvi.
46 Ibid., pp. 138–9.
47 Ibid., p. 45.
48 Ibid., p. 66.
49 Ibid., pp. 100, 103.
50 Ibid., pp. 112, 115.
51 Ibid., p. 165.
52 Mileson, *Parks*, p. 22.
53 Saunders, *Forest,* p. 9.
54 Mileson, *Parks*, pp. 99–100.
55 Ibid., p. 181.
56 Birrell, 'Records', p. 173.
57 Mileson, *Parks*, p. 24.

58 Ibid., p. 25.

59 Chris Wickham, 'European Forests in the Early Middle Ages: Landscape and Land Clearance', in *Land and Power: Studies in Italian and European Social History, 400–1200* (London, 1994), p. 160.

❧ THREE: Tree Movements

1 Andrew Coleby, 'Compton, Henry (1631/2–1713)', *Oxford Dictionary of National Biography* (Oxford, 2004); Sandra Morris, 'Legacy of a Bishop: The Trees and Shrubs of Fulham Palace Gardens Introduced 1675–1713', *Garden History*, 19 (1991), pp. 47–59, 49.

2 James Britten, 'Banister, John (1650–1692)', *Oxford Dictionary of National Biography*; Coleby, 'Compton'; Morris, 'Legacy', p. 49.

3 P. J. Jarvis, 'Plant Introductions to England and their Role in Horticultural and Silvicultural Innovation, 1500–1900', in *Change in the Countryside: Essays on Rural England, 1500–1900*, ed. H.S.A. Fox and R. A. Butlin (London, 1979), pp. 145–164, 153.

4 Mark Catesby, *Hortus Britanno-Americanus* (London, 1763).

5 John Claudius Loudon, *Arboretum et Fruticetum Britannicum* (London, 1838), vol. I, pp. 41, 45; John Ray, *Historia Plantarum* (London, 1686), II, pp. 1798–9; Britten, 'Banister'.

6 Douglas Chambers, *The Planters of the English Landscape Garden: Botany, Trees and the Georgics* (New Haven, CT, 1993), p. 3.

7 Loudon, *Arboretum*, p. 41; Chambers, *Planters*, pp. 36, 45; see Beryl Hartley, 'Exploring and Communicating Knowledge of Trees in the Early Royal Society', *Notes and Records of the Royal Society*, 64 (2010), pp. 229–50.

8 Loudon, *Arboretum*, p. 61.

9 Chambers, *Planters*, p. 111, quoting S. Switzer, *The Practical Husbandman* (London, 1733), vol. I, part I, p. liv.

10 H. Le Rougetel, 'Miller, Philip (1691–1771)', *Oxford Dictionary of National Biography*.

11 Loudon, *Arboretum*, vol. I, p. 54.

12 A. Murdoch, 'Campbell, Archibald, third duke of Argyll (1682–1761)', *Oxford Dictionary of National Biography*.

13 See M. Symes, A. Hodges and J. Harvey, 'The Plantings at Whitton', *Garden History*, 14 (1986), pp. 138–72.

14 Chambers, *Planters*, p. 92, quoting BL Add. MS 28727, fol. 5, 16 February 1747/48.

15 J. J. Cartwright (ed.), 'The Travels through England of Dr Richard Pococke Successively Bishop of Meath and of Ossory, during 1750, 1751, and Later Years', *Camden New Series*, 94 (1889), vol. II, pp. 260–61.

16 Chambers, *Planters*, p. 112, quoting a letter of 1 September 1741 from P. Collinson (1694–1768) to the American botanist and explorer J. Bartram (1699–1777). See also Douglas Chambers, 'Collinson, Peter (1694–1768)', *Oxford Dictionary of National Biography*.

17 M. Symes and J. H. Harvey, 'Lord Petre's Legacy: The Nurseries at Thorndon', *Garden History*, 24 (1996), pp. 272–82.

18 Nuala C. Johnson, 'Names, Labels and Planting Regimes: Regulating Trees at Glasnevin Botanic Gardens, Dublin, 1795–1850', in *Garden History*, 35 (2007), pp. 53–70.

19 See Paul A. Elliott, Charles Watkins and Stephen Daniels, *The British Arboretum: Trees, Science and Culture in the Nineteenth Century* (London, 2011).

20 George Sinclair, *Useful and Ornamental Planting* (London, 1832), p. 129.

21 George Nicholson, *The Illustrated Dictionary of Gardening: A Practical and Scientific Encyclopaedia of Horticulture* (London, 1889), vol. IV, pp. 450–57; Charles Sargent, *A Manual of the Trees of North America* (New York, 1905), preface.

22 William J. Bean, *Trees and Shrubs Hardy in the British Isles*, 7th edn (London, 1951) vol. I, preface, p. vii; Richard Drayton, *Nature's Government: Science, Imperial Britain and the 'Improvement' of the World* (New Haven, CT, 2001), pp. 221–68.

23 George Gordon, *The Pinetum*, 2nd edn (London, 1875), pp. 414–16; Bean, *Trees,* vol. III, pp. 303–5.

24 Elliott, Watkins and Daniels, *Arboretum*, pp. 172–3.

25 Ibid., pp. 69–70.

26 Printed Reports of the Garden Committee, numbers 1–5 (1823–7); Manuscript Minutes of the Gardening Committee (1818–1830), Lindley Library of the Royal Horticultural Society, London, M12/01; John Claudius Loudon, *Encyclopaedia of Gardening*, 5th edn (London 1830), pp. 1059–1060; Brent Elliott, *The History of the Royal Horticultural Society: 1804–2004* (London, 2004).

27 Loudon, *Arboretum,* vol. I, pp. 129–30; *Gardeners' Magazine*, 5 (1830), p. 346, fig. 79; 6 (1830), p. 250, fig. 44.

28 Loudon, *Encyclopaedia,* pp. 1059–60.

29 Elliott et al., *Arboretum,* p. 90.

30 Ibid., pp. 135–54; N. Jones, *Life and Death: Discourse on Occasion of the Lamented Death of Joseph Strutt* (London, 1844), p. 16.

31 J. R. Martin, 'Report on the State of Nottingham, Coventry, Leicester, Derby, Norwich and Portsmouth', in *Second Report of the Commissioners for Inquiry into the State of Large Towns and Populous Districts* (London, 1845), vol. II.

32 *Derby Mercury*, 20 August 1851.

33 C. M. Hovey, *Magazine of Horticulture*, 11 (1845), pp. 122–8.

34 Andrew Jackson Downing, *Rural Essays* (New York, 1856), pp. 497–557; Tom Schlereth, 'Early North-American Arboreta', *Garden History*, 35 (2007), pp. 196–216.

35 Evelyn Waugh, *Unconditional Surrender* (London, 1971), p. 91.

36 A. Jackson, 'Imagining Japan: The Victorian Perception of Japanese Culture', *Journal of Design History*, 5 (1996), pp. 245–56. See also Setsu Tachibana and Charles Watkins, 'Botanical Transculturation: Japanese and British Knowledge and Understanding of *Aucuba Japonica* and *Larix Leptolepis*, 1700–1920', *Environment and History*, 16 (2010), pp. 43–71.

37 Setsu Tachibana et al., 'Japanese Gardens in Edwardian Britain: Landscape and Transculturation', *Journal of Historical Geography*, 30 (2004), pp. 364–94.

38 E. Kaempfer (1651–1716), German traveller and naturalist; Carl Peter Thunberg (1743–1828), Swedish botanist; P. F. von Siebold (1796–1866), German doctor and botanist.

39 N. Kato, *Makino Hyohonkan shozou no Siebold Collection* (Siebold Collection at Makino Herbarium) (Kyoto, 2003); N. Kato, H. Kato, A. Kihara and M. Wakabayashi, *Makino Hyohonkan shozou no Siebold Collection*, CD Database (Tokyo, 2005).

40 Conrad Totman, *The Green Archipelago: Forestry in Preindustrial Japan* (Berkeley, CA, 1989), p. 261.

41 R. Ono and Y. Shimada, *Kai* (1763). Siebold used this book as a reliable and practical encyclopaedia and he collated and referenced his newly collected Japanese plants against the descriptions given in *Kai*.

42 B. Lindquist, 'Provenances and Type Variation in Natural Stands of Japanese Larch', *Acta Horti Gotoburgensis*, 20 (1955), pp. 1–34, 5.

43 H. J. Elwes and A. Henry, *The Trees of Great Britain and Ireland* (Edinburgh, 1907), vol. II, p. 384.

44 *The Gardeners' Chronicle*, 15 December 1860, p. 1103.

45 Rutherford Alcock, 'Narrative of a Journey in the Interior of Japan, Ascent of Fusiyama, and Visit to the Hot Sulphur Baths of Atami in 1860', read 13 May 1861, *Journal of the Royal Geographical Society*, 31 (1861), pp. 321–55; Rutherford Alcock, *The Capital of the Tycoon* (London, 1863), p. 483.

46 Elwes and Henry, *Trees*, vol. II, p. 385.

47 *Larix leptolepis*, Endlicher (1847); *Larix japonica*, Carrière (1855); *Larix kaempferi*, Sargent (1898); *Pinus larix*, Thunberg (1784); *Pinus kaempferi*, Lambert (1824); *Abies kaempferi*, Lindley (1833); *Abies leptolepis,* Siebold et Zuccarini (1842); *Pinus leptolepis*, Endlicher (1847).

48 Elwes and Henry, *Trees*, vol. II, p. 384.

49 Ibid., pp. 385–6.

50 Messrs. Dickson of Chester were said to have sold 750,000 trees in 1905. Elwes and Henry, *Trees*, vol. II, pp. 386–7.

51 J. MacDonald et al., 'Exotic Forest Trees in Great Britain: Paper Prepared for the Seventh British Commonwealth Forestry Conference, Australia and New Zealand', *Forestry Commission Bulletin*, 30 (London, 1957), p. 69.

52 Steve Lee, *Breeding Hybrid Larch in Britain*, Forestry Commission Information Note (2003); C. R. Lane, P. A. Beales, K.J.D. Hughes, R. L. Griffin, D. Munro, C. M. Brasier and J. F. Webber, 'First Outbreak of *Phytophthora ramorum* in England, on *Viburnum tinus*', *New Disease Reports*, 6 (2002), p.13; Forestry Commission, '*Phytophthora ramorum* in Larch Trees', www.forestry.gov.uk, 30 July 2012.

53 Forestry Commission, '*Phytophthora*'.

FOUR: Tree Aesthetics

1 Patrick Brydone, *A Tour through Sicily and Malta,* 1st edn 1773 (London, 1806), pp. 62–5; Katherine Turner, 'Brydone, Patrick (1736–1818)', *Oxford Dictionary of National Biography* (Oxford, 2004); G. E. Ortolani, *Biografia degli uomini illuustri della Sicilia* (Naples, 1818–21), vol. II.

2 Andrew Ballantyne, *Architecture, Landscape and Literature* (Cambridge, 1997). Knight intended to publish a description of his 'Expedition into Sicily' of 1777. A version was eventually published in translation by Goethe in 1810, and the original English version was rediscovered by Claudia Stumpf at Weimar in 1980 and published in 1986. See Richard Payne Knight, *Expedition into Sicily* [1777], ed. Claudia Stumpf (London, 1986).

3 Knight, *Expedition*, p. 58, Aci Reale 1 June 1777.

4 Castagno dei Cento Cavalli, Jean-Pierre Houël. 1776–1779 *Voyage pittoresque des Isles de Sicile, de Malte et de Lipari* (Paris, 1782).

5 William Linnard, *Welsh Woods and Forests* (Llandysul, 2000), Peniarth MS 28, p. 21.

6 Michael Camille, 'The "Très Riches Heures": An Illuminated Manuscript in the Age of Mechanical Reproduction', *Critical Inquiry*, 17 (1990), pp. 72–107.

7 G. Bartrum, *Dürer and His Legacy: The Graphic Work of a Renaissance Artist* (London, 2002).

8 Erik Hinterding, Ger Luijten and Martin Royalton-Kisch, *Rembrandt the Printmaker* (London, 2000), pp. 247–50.

9 Cynthia Schnieder, *Rembrandt's Landscapes* (New Haven, CT, and London, 1990); Christiaan Vogelaar and Gregor Weber, *Rembrandt's Landscapes* (Zwolle, 2006).

10 Seymour Slive, *Jacob van Ruisdael: A Complete Catalogue of His Paintings, Drawings and Etchings* (New Haven, CT, and London, 2001) p. 249; Seymour Slive, *Jacob van Ruisdael: Master of Landscape* (London, 2005), p. 46.

11 Peter Ashton, Alice I. Davies and Seymour Slive, 'Jacob van Ruisdael's Trees', *Arnoldia*, 42 (1982), pp. 2–31; Slive, *Complete Catalogue*, p. 268; Slive, *Ruisdael Landscape,* p. 29.

12 Michael Rosenthal, *The Art of Thomas Gainsborough* (New Haven, CT, and London, 1999), pp. 183–6; Slive, *Ruisdael Landscape*, p. 24.

13 Susan Sloman, *Gainsborough in Bath* (New Haven, CT, and London, 2002).

14 Michael Rosenthal and Martin Myrone, *Gainsborough* (London, 2003), pp. 78, 80.

15 Ibid., pp. 84, 96, 104; Elise L. Smith, '"The Aged Pollard's Shade": Gainsborough's Landscape with Woodcutter and Milkmaid', *Eighteenth-century Studies*, 41 (2007), pp. 17–39.

16 Helen Langdon, *Salvator Rosa* (London, 2010), p. 258; Charles Watkins and Ben Cowell, 'Letters of Uvedale Price', *Walpole Society*, 68 (2006), pp. 1–359, Uvedale Price to Sir George Beaumont, 12 January 1822, pp. 299–300.

17 David Watkin, *The English Vision* (London, 1982), p. vii.

18 William Hogarth, *Analysis of Beauty* (London, 1754), p. 52.

19 Isabel Chace, *Horace Walpole: Gardenist, an Edition of Walpole's The History of the Modern Taste in Gardening with an Estimate of Walpole's Contribution to Landscape Architecture* (Princeton, NJ, 1943), pp. 35–6.

20 Chace, *Walpole,* pp. 25–7, 31, 35–7.

21 William Gilpin, *An Essay upon Prints* (London, 1768), pp. 2–3.

22 William Gilpin, *Observations on the Western Parts of England Relative Chiefly to Picturesque Beauty* (London, 1798), p. 328.

23 William Gilpin, *Observations on the River Wye* (London, 1782), p. 1.

24 Uvedale Price, *Essays* (London, 1810), vol. i, p. 40.

25 Ibid., pp. 345, 22–3, 50, 114, 22, 55, 57, 244.

26 William Gilpin to William Mason, 12 July 1755; Carl P. Barbier, *William Gilpin: His Drawings, Teaching and Theory of the Picturesque* (Oxford, 1963), p. 53.

27 Robert Mayhew, 'William Gilpin and the Latitudinarian Picturesque', *Eighteenth-century Studies*, 33 (2000), pp. 349–66, 351.

28 John Ray, *The Wisdom of God Manifested in the Works of the Creation*, 1st edn 1691 (London, 1771), p. 87.

29 William Gilpin, *Remarks on Forest Scenery* (London, 1791), vol. i, p. 103, note.

30 Ibid., pp. 1–3.

31 Ibid., vol. ii, pp. 305–7.

32 Ibid., vol. i, pp. 7–8; John Considine, 'Lawson, William (1553/4–1635)', *Oxford Dictionary of National Biography*, (Oxford, 2004); William Lawson, *A New Orchard and Garden, or, The Best Way for Planting, Grafting, and to Make Any Ground Good for a Rich Orchard; Particularly in the North Parts of England* (London, 1618).

33 Ibid., vol. i, pp. 8–9, 14.

34 Ibid., vol. i, pp. 10–14; Price, *Essays*, vol. i, p. 244.

35 Watkins and Cowell, 'Letters', Uvedale Price to Lord Abercorn, 31 May 1796, p. 85.

36 Ibid., Uvedale Price to Lord Aberdeen, 6 February 1818, p. 274. Price may have come across this analogy via Joseph Addison, who in *The Spectator*, 215, 6 November 1706, stated that Aristotle in his *Doctrine of Substantial Forms* tells us that 'a Statue lies hid in a Block of Marble'.

37 Ibid., Uvedale Price to Lady Beaumont, August 1803, p. 165.

38 Gilpin, *Forest Scenery*, vol. i, pp. 4, 34, 15.

39 Ibid., p. 4; Arthur Young, *A Six Weeks Tour through the Southern Counties of England and Wales* (London, 1769) pp. 92, 308; J. Middleton, *General View of the Agriculture of the County of Middlesex* (London, 1813), pp. 344–7; Watkins and Cowell, 'Letters', Uvedale Price to Lord Abercorn, 14 July 1792, p. 79.

40 Linnard, *Welsh Woods*, p. 156; H. S. Steuart, *The Planter's Guide: or A Practical Essay on the Best Method of Giving Immediate Effect to Wood by the Removal of Large Trees and Underwood* (Edinburgh, 1828), p. 60; J. G. Strutt, *Sylva Britannica; or, Portraits of Forest Trees, Distinguished for their Antiquity, Magnitude, or Beauty* (London, 1830), p. 98; J. Main, *The Forest Planter and Pruner's Assistant, Being a Practical Treatise on the Management of the Native and Exotic Forest Trees Commonly Cultivated in Great Britain* (London, 1839), p. 236.

41 W. M. Craig, *A Course of Lectures on Drawing, Painting, and Engraving Considered as Branches of Elegant Education Delivered in the Saloon of the Royal Institution* (London, 1821), p. 286; Anonymous, *Woodland Gleanings: An Account of British Forest-trees* (London, 1865), p. 20; Anonymous, *English Forests and Forest Trees: Historical, Legendary, and Descriptive* (London, 1853), p. 107; J. Dagley and P. Burman, 'The Management

of Pollards of Epping Forest: Its History and Revival' in *Pollard and Veteran Tree Management II*, ed. Helen Read (London, 1996), pp. 29–41.

❧ FIVE: Pollards

1 Sandrine Petit and Charles Watkins, 'Pollarding Trees: Changing Attitudes to a Traditional Land Management Practice in Britain, 1600–1900', *Rural History*, 14 (2003), pp. 157–76; Helen Read, *Veteran Trees: A Guide to Good Management* (Peterborough, 2000).

2 Frans Vera, *Grazing Ecology and Forest History* (Wallingford, 2000).

3 I. Austad, 'Tree Pollarding in Western Norway', in *The Cultural Landscape: Past, Future, Present*, ed. H. Birks et al. (Cambridge, 1988), pp. 11–29; E. Bargioni and A. Z. Sulli, 'The Production of Fodder Trees in Valdagano, Vicenza, Italy', in *The Ecological History of European Forests*, ed. Keith J. Kirby and Charles Watkins (Wallingford, 1998), pp. 43–52; C.-A. Hæggström, 'Pollard Meadows: Multiple Use of Human-made Nature', in *The Ecological History of European Forests*, ed. Kirby and Watkins, pp. 33–42; Paul Halstead, 'Ask the Fellows Who Lop the Hay: Leaf-fodder in the Mountains of Northwest Greece', *Rural History*, 9 (1998), pp. 211–34; F.-X. Trivière, 'Emonder Les Arbres: Tradition Paysanne, Pratique Ouvrière', *Terrain*, March (1991), pp. 62–77.

4 Sandrine Petit, 'Parklands with Fodder Trees: a Full Response to Environmental and Social Changes', *Applied Geography*, 23 (2003), pp. 205–25; J. Anderson et al., 'Le Fourrage Arboré à Bamako: Production et Gestion des Arbres Fourragers, Consommation et Filières d'Approvisionnement', *Sécheresse*, 2 (1994) pp. 99–105; Food and Agriculture Organization, *Forests, Trees and Food* (Rome, 1992).

5 R. C. Khanal and D. B. Subba, 'Nutritional Evaluation of Leaves from Some Major Fodder Trees Cultivated in the Hills of Nepal', *Animal and Feed Science Technology*, 92 (2001), pp. 17–32; D. A. Gilmour and M. C. Nurse, 'Farmer Initiatives in Increasing Tree Cover in Central Nepal', *Mountain Research and Development*, 11 (1991), pp. 329–37; J. L. Hellin et al., 'The Quezungual System: An Indigenous Agroforestry System from Western Honduras', *Agroforestry Systems*, 46 (1999), pp. 229–37.

6 Eirini Saratsi, 'Landscape History and Traditional Management Practices in the Pindos Mountains, Northwest Greece, *c*. 1850–2000', PhD thesis (University of Nottingham, 2003), pp. 199–207.

7 Ibid.

8 Pantelis Arvanitis, 'Traditional Forest Management in Psiloritis, Crete, *c*. 1850–2011: Integrating Archives, Oral History and GIS', PhD thesis (University of Nottingham, 2011), pp. 203–55.

9 Ibid.

10 Ruth M. Tittensor, 'A History of the Mens: A Sussex Woodland Common', *Sussex Archaeological Collections*, 116 (1977–8), pp. 347–74; Oliver Rackham, *The Last Forest: The Story of Hatfield Forest* (London, 1989), p. 247.

11 Petit and Watkins, 'Pollarding Trees'; N.D.G. James, *An Historical Dictionary of Forestry and Woodland Terms* (Oxford, 1991); Richard Muir,

'Pollards in Nidderdale: A Landscape History', *Rural History*, 11 (2001), pp. 95–111.

12 J. Worlidge, *A Compleat System of Husbandry and Gardening* (London, 1669), p. 133, in the 1728 edition it is p. 212; Moses Cook, *The Manner of Raising, Ordering, and Improving Forest-trees: With Directions How to Plant, Make, and Keep Woods, Walks, Avenues, Lawns, Hedges, &c* (London, 1676), pp. 42, 141; Batty Langley, *A Sure Method of Improving Estates by Plantations of Oak, Elm, Ash, Beech, and Other Timber-trees, Coppice-woods, &c.* (London, 1728), p. 212.

13 Arthur Standish, *New Directions of Experience Authorized by the Kings Most Excellent Majesty, as May Appear, for the Increasing of Timber and Fire-wood with the Least Waste and Losse of Ground* (London, 1615), p. 21, 9; Arthur Standish, *The Commons Complaint* (London, 1611), p. 9; Worlidge, *Husbandry*, p. 126; John Mortimer, *The Whole Art of Husbandry: Or the Way of Managing and Improving of Land* (London, 1707), pp. 331, 393.

14 John Evelyn, *Sylva*, 1st edn (London, 1664), p. 213; 2nd edn (1670), p. 142; 4th edn (1706), p. 208; 2nd edn (1670), p. 141; 1st edn (1664), p. 19; 4th edn (1706), pp. 46–7.

15 Cook, *Forest-trees*, pp. 18, 110, 102; W. Ellis, *The Timber-tree Improved or the Best Practical Methods of Improving Different Lands with Proper Timber* (London, 1742), p. 106.

16 Arthur Young, 'French Edict in Consequence of the Scarcity in France', in *Annals of Agriculture and Other Useful Arts* (London, 1785), pp. 63–71, 62; William Marshall, *Planting and Rural Ornament*, 2nd edn (London, 1796), pp. 100, 142, 182.

17 John Middleton, *General View of the Agriculture of the County of Middlesex* (London, 1813), p. 345; W. T. Pomeroy, *General View of the Agriculture of the County of Worcester* (London, 1794), p. 21; J. Clark, *General View of the Agriculture of the County of Hereford* (London, 1794), p. 66. See also John Barrell, *The Idea of Landscape and the Sense of Place, 1730–1840* (Cambridge, 1972).

18 W. Pearce, *General View of the Agriculture of the County of Berkshire* (London, 1794), p. 57; T. Stone, *General View of the Agriculture of the County of Bedford* (London, 1794), p. 53; J. Clark, *General View of the Agriculture of the County of Radnor* (London, 1794), p. 28.

19 J. Anderson, *Essays Relating to Agriculture and Rural Affairs* (Edinburgh, 1784), vol. II, p. 17, fn.

20 Uvedale Price, 'On the Bad Effects of Stripping and Cropping Trees', *Annals of Agriculture*, 5 (1786), pp. 241–3.

21 Price, 'Stripping', pp. 247–9.

22 Herefordshire Record Office, BC 986A.

23 J. Priest, *General View of the Agriculture of the County of Buckinghamshire* (London, 1813), pp. 258–9; Middleton, *Middlesex*, pp. 344–7; Clark, *Hereford*, p. 26.

24 Christopher Hibbert, *Queen Victoria in Her Letters and Journals: A Selection* (London, 1985), p. 273.

25 Elizabeth Baigent, 'A "Splendid Pleasure Ground [for] the Elevation and Refinement of the People of London": an Historical Geography of Epping

Forest, 1860–95', in *English Geographies, 1600–1950: Historical Essays on English Customs, Cultures and Communities in Honour of Jack Langton*, ed. Elizabeth Baigent and Robert Mayhew (Oxford, 2009), pp. 104–26; Edward Buxton, *Epping Forest* (London, 1884).

SIX: Sherwood Forest

1 *Ivanhoe* was first published on 18 December in Edinburgh and 31 December 1819 in London, 'so close to the end of the year, *Ivanhoe* bore the date 1820 on its title-page', Walter Scott Digital Archive, University of Edinburgh, www.walterscott.lib.ed.ac.uk.

2 Mark Girouard, *The Return to Camelot: Chivalry and the English Gentleman* (London, 1981).

3 Sir Walter Scott, *Ivanhoe* (Edinburgh, 1820), pp. 195, 561.

4 Washington Irving, *Abbotsford and Newstead Abbey* (London, 1835), pp. 233–4.

5 There is a large literature on the history of royal forests. See for example: Charles R. Young, *The Royal Forests of Medieval England* (Philadelphia, PA, 1979); Oliver Rackham, *Ancient Woodland* (London, 1980); Rackham, *The Last Forest: The Story of Hatfield Forest* (London, 1989); N.D.G. James, *A History of English Forestry* (London, 1981); John Langton and Graham Jones, *Forests and Chases of England and Wales, c. 1000 – c. 1500* (Oxford, 2010); Mary Wiltshire et al., *Duffield Frith: History and Evolution of a Medieval Derbyshire Forest* (Ashbourne, 2005).

6 Stephanos Mastoris and Sue Groves, *Sherwood Forest in 1609: A Crown Survey by Richard Bankes* (Nottingham, 1997).

7 Jacob George Strutt, 'Introduction', in *Sylva Britannica* (London, 1826).

8 J. C. Holt, *Robin Hood* (London, 1982).

9 Charles Watkins, '"A Solemn and Gloomy Umbrage": Changing Interpretations of the Ancient Oaks of Sherwood Forest', in *European Woods and Forests: Studies in Cultural History*, ed. Charles Watkins (Wallingford, 1998), pp. 93–114.

10 'The Fourteenth Report of the Commissioners Appointed to Enquire into the State and Condition of the Woods, Forests, and Land Revenues of the Crown, and to Sell or Alienate Fee Farm and Other Unimprovable Rents', *House of Commons Journal*, 48 (1793), p. 469.

11 Ibid., p. 469.

12 Ibid., pp. 473, 509.

13 Ibid., p. 481.

14 Ibid., pp. 473, 481.

15 Ibid., pp. 472–3.

16 Ann Gore and George Carter, *Humphry Repton's Memoirs* (Norwich, 2005), p. 106; Hayman Rooke, *A Sketch of the Ancient and Present Extent of Sherwood Forest, in the County of Nottingham* (Nottingham, 1799), incorporated material from the Commissioners Report on Sherwood Forest of 1791 and the results of some of his archaeological work.

17 Rooke, *Sketch*, pp. 5–6.

18 Evelyn, *Sylva* (1670), pp. 159–60.

19 Rooke, *Sketch*, pp. 16–17.

20 Ibid., pp. 18–19.

21 William Howitt, 'Sherwood Forest', in *The Rural Life of England* (London, 1838), pp. 383, 387, 385.

22 January Searle, *Leaves from Sherwood Forest* (London, 1850), pp. 71–2.

23 Christopher Thomson, *Autobiography of an Artisan* (London, 1847), pp. 301–2.

24 Christopher Thomson, *Hallamshire Scrapbook*, 4.

25 Anonymous, *Worksop, 'The Dukery' and Sherwood Forest* (London, 1850) p. 244; E. Eddison, *History of Worksop: With Historical, Descriptive and Discursive Sketches of Sherwood Forest* (London, 1854), pp. 194–5.

26 Ford Madox Ford, 'The Beautiful Genius Ivan Turgenev', in *Memories and Impressions* (London, 1971), p. 129.

27 F. Sissons, *Beauties of Sherwood Forest* (Worksop, 1888), p. 58.

28 Searle, *Leaves*, pp. 72, 104.

29 *Victoria County History of Nottinghamshire* (London, 1908), vol. I, p. 93.

30 Ibid., p. 100.

31 Ibid., p. 156.

32 W. J. Sterland, *The Birds of Sherwood Forest* (London, 1869), p. 144.

33 Sissons, *Sherwood Forest*, p. 2.

34 Nottingham University Manuscripts Department (NUMD) Ma 2S 7, p. 117.

35 Sissons, *Sherwood Forest*, p. 71.

36 Joseph Rodgers, *Sherwood Forest* (London, 1908), p. 16; Nottingham University Manuscripts Department (NUMD) MA 2C 132.

37 NUMD Ma 2C 133; NUMD Ma 2C 208.

38 NUMD Ma 4E 81.

39 NUMD Ma 3E 151.

40 James, *English Forestry*, p. 247.

41 NUMD Ma 5E 211.

42 NUMD Ma 5E 211.

43 NUMD Ma 4A 7/41.

44 NUMD Ma 4A 7/49/2.

45 NUMD Ma 4A 7/56.

46 NUMD Ma 4A 7/43.

47 NUMD Ma 3E 4799.

48 NUMD Ma 2A 94.

49 Interview with George Holt (1998).

50 Royal Forestry Society 1984.

51 NUMD Ma 4A 7/41.

52 Brian Wood, 'Land Management and Farm Structure: Spatial Organisation on a Nottinghamshire Landed Estate and its Successors, 1860–1978', PhD Thesis (University of Nottingham, 1981).

53 Interview with John Irbe (1998).

54 NUMD Ma 4A 7/41.

55 Nikolaus Pevsner, *Nottinghamshire* (London, 1979), p. 120. Architects Ian Pryer, William Saunders and Partners 1973–6.

56 NUMD Ma 4A 7/41.

57 C. R. McLeod et al., *The Habitats Directive: Selection of Special Areas of Conservation in the UK*, 2nd edn (Peterborough, 2005), www.jncc.gov.uk/sacselection; Charles Watkins et al., 'The Use of Dendrochronology to Evaluate Dead Wood Habitats and Management Priorities for the Ancient Oaks of Sherwood Forest', in *Forest Biodiversity: Lessons from History for Conservation*, ed. O. Honnay et al. (Wallingford, 2004), pp. 247–67.

58 Interview with Andrew Poole (1998).

SEVEN: Estate Forestry

1 John Stoddart, *Remarks on the Scenery and Manners in Scotland during the Years 1799 and 1800* (Edinburgh, 1801). The drawing is by Hugh William 'Grecian' Williams, BM 1872,0413.289.

2 James Boswell, *The Life of Samuel Johnson* (Edinburgh, 1823), vol. II, p. 316.

3 Jane Austen, *Mansfield Park* (London, 1814), chapter 6.

4 D. Hudson and K. W. Luckhurst, *The Royal Society of Arts: 1754–1954* (London, 1954), pp. 86–9.

5 Uvedale Price, *Essays* (London, 1810), vol. I, pp. 259–63.

6 J. C. Loudon, *Observations on the Formation and Management of Useful and Ornamental Plantations* (Edinburgh, 1804).

7 John Claudius Loudon, *Arboretum et Fruticetum Britannicum* (London, 1838), vol. I, pp. 1–2.

8 Stephen Daniels and Charles Watkins, 'Picturesque Landscaping and Estate Management: Uvedale Price at Foxley, 1770–1829', *Rural History*, 2 (1991), pp. 141–70; J. M. Neeson, *Commoners: Common Right, Enclosure and Social Change in England, 1700–1820* (Cambridge, 1993). See also Nicola Whyte, 'An Archaeology of Natural Places: Trees in the Early Modern Landscape', *Huntington Library Quarterly*, 76 (2013), pp. 499–517; Carl Griffin, 'Protest Practice and (Tree) Cultures of Conflict: Understanding the Spaces of "Tree Maiming" in Eighteenth and Early Nineteenth-century England', *Transactions of the Institute of British Geographers*, 33 (2008), pp. 91–108.

9 Tim Shakesheff, 'Wood and Crop Theft in Rural Herefordshire, 1800–60', *Rural History*, 13 (2002), pp. 1–17, 14.

10 E.J.T. Collins, 'Woodlands and Woodland Industries in Great Britain during and after the Charcoal Iron Era', in *Protoindustries et Histoire des Forêts*, ed J.-P. Métaillé (Toulouse, 1992), pp. 109–20.

11 J. Main, *The Forest Planter and Pruner's Assistant* (London, 1839); J. West, *Remarks on the Management or Rather the Mismanagement of Woods, Plantations and Hedgerow Timber* (London, 1842); J. Standish and C. Noble, *Practical Hints on Planting Ornamental Trees* (London, 1852).

12 William Ablett, *English Trees and Tree Planting* (London, 1880), p. 402.

13 Tom Bright, *Pole Plantations and Underwoods* (London, 1888), pp. 20–21.

14 J. Nisbet, *James Brown's The Forester*, 6th edn (London, 1894), vol. I, p. 44.

15 J. Nisbet, *The Forester* (London, 1905), vol. I, p. 49.

16 H. FitzRandolph and M. Hay, *The Rural Industries of England and Wales* (Oxford, 1926); Herbert Edlin, *Woodland Crafts in Britain* (London, 1949); Eastnor Estate coppice sale records, Eastnor Estate Muniments, 1933.

17 George Sinclair, *Useful and Ornamental Planting* (London, 1832), p. 2.

18 Eric Richards, *The Highland Clearances* (Edinburgh, 2000).

19 J.L.F. Fergusson, 'Forestry in Perthshire: Notes on Past History', *Forestry*, 29 (1956), pp. 84–5.

20 Ibid., p. 86.

21 H. M. Steven, 'Silviculture of Conifers in Britain', *Forestry* (1927), p. 9.

22 K. Garlick and A. Macintyre, '3 October 1801', in *The Diary of Joseph Farington* (New Haven, CT, and London, 1978–98), p. 1644.

23 Richard Moore-Colyer, 'Thomas Johnes (1748–1816)', *Oxford Dictionary of National Biography* (Oxford, 2004); William Linnard, *Welsh Woods and Forests* (Llandysul, 2000), Peniarth MS 28.

24 Price, *Essays*, vol. I, p. 26.

25 Ibid., pp. 265–6.

26 Ibid., pp. 26–7.

27 Ibid., pp. 273–6.

28 Charles Watkins and Ben Cowell, 'Letters of Uvedale Price', *Walpole Society*, 68 (2006).

29 Price, *Essays*, vol. I, pp. 270–73.

30 John Claudius Loudon, *Hints on the Formation of Gardens and Pleasure-grounds with Designs in Various Styles of Rural Embellishment* (London, 1812).

31 John Claudius Loudon, *Encyclopaedia of Gardening* (London, 1830), p. 943.

32 Sophieke Piebenga, 'William Sawrey Gilpin', *Oxford Dictionary of National Biography* (Oxford, 2004).

33 R. Monteath, *The Forester's Guide and Profitable Planter,* 2nd edn (Edinburgh, 1824); A. C. Forbes, *English Estate Forestry* (London, 1904), pp. 16–17.

34 Linnard, *Welsh Woods,* pp. 160–61.

35 Forbes, *English Forestry*, p. 318.

36 Ibid.

37 J. Brown, *The Forester,* 4th edn (Edinburgh, 1871), advertisement, p. 836.

38 Forbes, *English Forestry*, p. 17.

39 Brown, *Forester*, p. 568.

40 J. Brown, *The Forester* (Edinburgh, 1851), pp. 138, 420.

41 Loudon, *Plantations*; Jane Loudon, *A Short Account of the Life and Writings of John Claudius Loudon* (London, 1845), pp. xiii–xvi.

42 Paul Elliott et al., 'William Barron (1805–91) and Nineteenth-century British Arboriculture: Evergreens in Victorian Industrializing Society', *Garden History*, 35 supplement 2 (2008), pp. 129–48.

43 William Barron, *The British Winter Garden* (London, 1853), pp. 9–24.

44 'Champion Trees', at www.bicton.ac.uk.

45 Thomas Baines, 'Eastnor Castle, Ledbury: The Seat of Earl Somers', *Gardeners' Chronicle*, 19 January (1878), p. 76.

46 William Gilpin, *Observations on the River Wye . . .* 2nd edn (London, 1789), p. 6.

47 Colin Ellis, *Leicestershire and the Quorn Hunt* (Leicester, 1951), p. 61.

48 J. Otho Paget, *Hunting* (London, 1900), pp. 81–4.

49 C. C. Rogers, 'Pheasant Management and Shooting in Hill Countries', in *Shooting*, ed. Horace G. Hutchinson (London, 1903), p. 55.

50 Patrick Waddington, *Turgenev and England* (London, 1980), pp. 15, 59, 228–9.

51 Ibid., pp. 228–9.

52 Alexander J. Napier, 'Pheasants at Holkham', in *Shooting*, ed. Horace G. Hutchinson (London, 1903), pp. 35–6.

53 John Simpson, *Game and Game Coverts* (Sheffield, 1907).

54 Rogers, 'Pheasant Management', pp. 59–64.

55 C. J. Cornish, 'Pheasants at Nuneham', in *Shooting*, ed. Hutchinson, p. 45.

56 Jonathan Garnier Ruffer, *The Big Shots* (London, 1977), pp. 79–81.

57 J. Ruskin, *Modern Painters*, vol. v (1860), in *Selections from the Writings of John Ruskin* (Edinburgh, 1907), pp. 79–81.

58 John Ruskin to John Davidson, 24 February 1887, in *Transactions of the English Arboricultural Society* (1887), pp. 156–7.

EIGHT: Scientific Forestry

1 Hannss Carl von Carlowitz, *Sylvicultura Oeconomica* (Leipzig, 1713); Christoph Ernst 'An Ecological Revolution? The "Schlagwaldwirtschaft" in Western Germany in the Eighteenth and Nineteenth Centuries', in *European Woods and Forests: Studies in Cultural History*, ed. Charles Watkins (Wallingford, 1998), pp. 83–92.

2 Ibid., pp. 84, 90.

3 Richard Keyser, 'The Transformation of Traditional Woodland Management: Commercial Sylviculture in Medieval Champagne', *French Historical Studies*, 32 (2009), pp. 353–84, 356, 361.

4 Ibid., pp. 366, 368.

5 Ibid., pp. 366, 368, 372, 375.

6 Ibid., pp. 379–80.

7 Ibid., pp. 356–7, 380, 384.

8 Paul Warde, *Ecology, Economy and State Formation in Early Modern Germany* (Cambridge, 2006), p. 171; Joachim Radkau, 'Wood and Forest in German History: In Quest of an Environmental Approach', *Environment and History*, 2 (1996), pp. 63–76, 65.

9 Radkau, 'Wood and Forest', p. 66.

10 Ibid., pp. 71–2.

11 Gregory Barton, 'Empire Forestry and the Origins of Environmentalism', *Journal of Historical Geography*, 27 (2001), pp. 529–52.

12 David Prain, 'Brandis, Sir Dietrich (1824–1907)', revd M. Rangarajan, *Oxford Dictionary of National Biography* (Oxford, 2004); R. S. Troup, 'Schlich, Sir William Philipp Daniel (1840–1925)', revd Andrew Grout, *Oxford Dictionary of National Biography* (Oxford, 2004).

13 William Schlich, *A Manual of Forestry* (London, 1889–95) vol. i, pp. v–vii.

14 Troup, 'Schlich'.

15 Schlich, *Forestry*, vol. iii, p. ix.

16 Dr Richard Hess, Professor of Forestry, University of Geissen; Dr Karl
   Gayer, Professor of Forestry, University of Munich.

17 John Nisbet, *Studies in Forestry* (Oxford, 1894), pp. vii–viii.

18 J. M. Powell, '"Dominion over Palm and Pine": The British Empire
   Forestry Conferences, 1920–47', *Journal of Historical Geography*, 33 (2007),
   pp. 852–77.

19 Karen Hovde, 'Charles Sprague Sargent', *Forest History Today* (Spring
   2002), pp. 38–9; Barton, 'Empire Forestry'.

20 Charles Sprague Sargent, *Garden and Forest*, 9 (1896) pp. 191–2, quoted
   by Barton, 'Empire Forestry', p. 540.

21 Barton, 'Empire Forestry', pp. 541, 542; Gifford Pinchot, *Breaking New
   Ground* (Washington, DC, 1998), p. 27.

22 Carl Alwin Schenck Photograph Series MC35, NSCU Library, note on reverse
   of photograph of Frederick Law Olmsted, at www.lib.ncsu.edu; Carl Alwin
   Schenck, *Cradle of Forestry in America: The Biltmore Forest School,
   1898–1913* (Durham, NC, 2001).

23 A. B. Recknagel, *The Theory and Practice of Working Plans*, 2nd edn
   (New York, 1917), p. v.

24 John Simpson, *The New Forestry* (Sheffield, 1903).

25 A. C. Forbes, *The Development of British Forest* (1910), p. 252; N.D.G.
   James, *A History of English Forestry* (London, 1981); N.D.G. James, 'A
   History of Forestry and Monographic Forestry Literature in Germany,
   France and the United Kingdom', in *The Literature of Forestry and
   Agroforestry*, ed. P. McDonald and J. Lassoie (New York, 1996), pp. 15–44;
   Judith Tsouvalis and Charles Watkins, 'Imagining and Creating Forests in
   Britain 1890–1939,' in *Forest History: International Studies on Socio-
   Economic and Forest Ecosystem Change*, ed. Mauro Agnoletti and
   S. Anderson (Wallingford, 2000), pp. 371–86.

26 James, *English Forestry*, p. 75.

27 Thomas Bewick, *Quarterly Journal of Forestry* (1914); William Schlich,
   'Report on the Visit of Royal English Arboricultural Society to German
   Forests', *Quarterly Journal of Forestry*, 8 (1914), pp. 75–81.

28 David E. Evans, 'Robinson, Roy Lister, Baron Robinson
   (1883–1952)', *Oxford Dictionary of National Biography* (Oxford, 2004);
   G. B. Ryle, *Forest Service: The First Forty-five Years of the Forestry
   Commission of Great Britain* (Newton Abbot, 1969), pp. 21–2.

29 Evans, 'Roy Robinson'; James, *English Forestry*, p. 196. Anon., 'Report on
   the REAS Meeting in September 1916 on the Present Position and Future
   Development of Forestry in England and Wales', *Quarterly Journal of
   Forestry*, 11 (1917), pp. 20–58.

30 John V. Beckett, *The Aristocracy in England: 1660–1914* (Oxford, 1986);
   David Cannadine, *The Decline and Fall of the British Aristocracy* (London,
   1996).

31 Myfanwy Piper, 'Nash, Paul (1889–1946)', revd Andrew Causey, *Oxford
   Dictionary of National Biography* (Oxford, 2004).

32 Andrew Causey, 'Introduction', in *Paul Nash: Paintings and Watercolours*
   (London, 1975).

33 Ryle, *Forest Service*, p. 23.

34 Ibid., pp. 26–8.
35 Ibid., pp. 43, 28.
36 Forestry Commission, *Forestry Practice: Forestry Commission Bulletin 14* (London, 1933); R. F. Wood and M. Nimmo, *Chalk Downland Afforestation: Forestry Commission Bulletin 34* (London, 1962), Foreword.

❧ NINE: Recreation and Conservation

1 Kim Novak interview with Stephen Rebello, March 2003, www.labyrinth.net.au.
2 Wilson E. Albee in *San Jose Mercury*, 22 April 1917, quoted by Eugene T. Sawyer, *Santa Clara County, California* (Los Angeles, CA, 1922), p. 206.
3 Big Basin Redwoods State Park California, 2011 Leaflet, at www.parks.ca.gov.
4 Dennis R., 'Dean Muir, John (1838–1914)', *Oxford Dictionary of National Biography* (Oxford, 2004); Donald Worster, *A Passion for Nature: The Life of John Muir* (Oxford, 2008).
5 Sue Shephard, 'Lobb, William (1809–1863)', *Oxford Dictionary of National Biography* Worster, *Passion*.
6 George Gordon, *The Pinetum*, 2nd edn (London, 1875), pp. 414–16; William J. Bean, *Trees and Shrubs Hardy in the British Isles*, 7th edn (London, 1951), vol. III, pp. 303–5.
7 National Gallery of Art, *Carleton Watkins: The Art of Perception*, www.nga.gov.
8 John Auwaerter and John Sears, *Historic Resource Study for Muir Woods National Monument* (Boston, 2006), p. 59.
9 Worster, *Passion*, p. 170.
10 John Muir, *Our National Parks* (Boston, 1901).
11 Ronald A. Bosco and Glen M. Johnson, *Journals and Miscellaneous Notebooks of Ralph Waldo Emerson* (Cambridge, MA, 1982) vol. XVI, 12 May 1871, pp. 237–9.
12 Muir, *National Parks*.
13 Auwaerter and Sears, *Muir Woods*, pp. 41–3.
14 Ibid., pp. 48–50.
15 F. E. Olmsted, *Muir National Monument Redwood Canyon: Marin County, California*, unpublished report, 26 September 1907, pp. 4–5, Muir Woods Park Files, quoted in Auwaerter and Sears, *Muir Woods*, p. 73.
16 William Kent, 'Redwoods', *Sierra Club Bulletin*, 6 (June 1908), pp. 286–7.
17 A. MacEwen and M. MacEwen, *National Parks: Conservation or Cosmetics?* (London, 1982).
18 The National Archives TNA FI8 162, Forestry Commission Mimeograph (1929).
19 Ibid.
20 George Revill and Charles Watkins, 'Educated Access: Interpreting Forestry Commission Forest Park Guides', in *Rights of Way: Policy, Culture and Management*, ed. Charles Watkins (London, 1996), pp. 100–128.
21 Forestry Commission, *Report of the National Forest Park Committee 1935* (London, 1935), pp. 2–6.

22 Internal report written by Sir Roy Robinson for Sir Francis Acland (Commissioner, 1919–39), on the Forestry Commissioners' contributions towards the idea underlying National Parks, *c.* 1936.

23 *Hansard*, 9 December 1936, p. 2105.

24 J. R. Aaron, 'H. L. Edlin, MBE', *Forestry*, 55 (1977), pp. 203–5.

25 H. L. Edlin, 'Fifty Years of Forest Parks', *Commonwealth Forestry Review*, 48 (1969), pp. 113–26, 125.

26 Forestry Commission, *Thirtieth Annual Report of the Forestry Commissioners for the Year Ending 30th September 1949* (London, 1950), p. 87.

27 Forestry Commission, *Thirty-second Annual Report of the Forestry Commissioners for the Year Ending 30 September 1951* (London, 1952), p. 50.

28 Forestry Commission, *National Forest*, p. 5.

29 Forestry Commission, *Report of the National Forest Park Committee (Forest of Dean)* (London, 1938), pp. 4–6.

30 W. E. S. Mutch, *Public Recreation in National Forests: A Factual Survey* (London, 1968), p. 83.

31 Edlin, 'Fifty Years', p. 125.

32 William Wordsworth, 'A Guide through the District of the Lakes' (1835), in *William Wordsworth: Selected Prose*, ed. J. Hayden (London, 1988), p. 57.

33 Miles Hadfield, *Landscape with Trees* (London, 1967), p. 181.

34 H. H. Symonds, *Afforestation in the Lake District: A Reply to the Forestry Commission's White Paper of 26 August 1936* (London, 1936), pp. 51, 16.

35 Edlin, 'Fifty Years', pp. 117, 116.

36 Alan Horne, 'Rooke, Noel (1881–1953)', *Oxford Dictionary of National Biography* (Oxford, 2004).

37 John Walton, *The Border Forest Park Guide* (London, 1962), p. 1.

38 John Moore, *Wood Properties and Uses of Sitka Spruce in Britain* (Edinburgh, 2011), pp. 1–2.

39 H. L. Edlin, 'Britain's New Forest Villages', *Canadian Forestry Gazette* (1953), pp. 151–8.

40 Frank E. Lutz, *Nature Trails: An Experiment in Outdoor Education*, Miscellaneous Publication 21, The American Museum of Natural History (New York, 1926).

41 Ralph H. Lewis, *Museum Curatorship in the National Park Service, 1904–1982*, National Park Service (Washington, DC, 1993), p. 38.

42 Lutz, *Nature Trails*, p. 3.

43 Ibid., p. 36.

44 J. N. Rogers, *Yosemite Nature Notes*, 9 March (1930), p. 17, at www.nps.gov.

45 David Matless et al., 'Nature Trails: The Production of Instructive Landscapes in Britain, 1960–72', *Rural History*, 21 (2010), pp. 97–131.

TEN: Ligurian Semi-natural Woodland

1 Diego Moreno, *Dal Documento: Storia e Archeologia dei Sistemi Agro-Sylvo-Pastorali* (Bologna, 1990); Ross Balzaretti et al., *Ligurian Landscapes: Studies in Archaeology, Geography and History* (London, 2004); Roberta Cevasco, *Memoria Verde: Nuovi Spazi per La Geografia* (Reggio Emilia, 2007).

2 ASP Uff. Confini, b.266/1. Paulo De Podio, public notary, 27/09/1564 (Reference provided by Roberta Cevasco).

3 Diego Moreno and Osvaldo Raggio, 'The Making and Fall of an Intensive Pastoral Land-use System: Eastern Liguria, 16–19th Centuries', *Rivista di Studi Liguri*, LVI (1990), pp. 193–217, 196.

4 Ibid., p. 204.

5 Archivio De Paoli, Porciorasco, N. 8, Museo Contadini, Cassego.

6 Archivio De Paoli, Porciorasco, N. 13, Museo Contadini, Cassego.

7 The coring was done by Roberta Cevasco, Diego Moreno and Charles Watkins; the ring counting was undertaken by Robert Howard, Nottingham Tree-ring Dating Laboratory.

8 M. Conedera et al., 'The Cultivation of Castanea Sativa (Mill.) in Europe, from its Origin to its Diffusion on a Continental Scale', *Vegetation History and Archaeobotany*, 13 (2004), pp. 161–79, 165.

9 Nick P. Branch, 'Late Würm Lateglacial and Holocene Environmental History of the Ligurian Apennines, Italy', in *Ligurian Landscapes: Studies in Archaeology, Geography and History*, ed Ross Balzaretti et al. (London, 2004), p. 57.

10 Conedera et al., p. 61.

11 Ross Balzaretti, *Dark Age Liguria* (London, 2013), p. 57.

12 Enzo Baraldi et al., 'Ironworks Economy and Woodmanship Practices, Chestnut Woodland Culture in Ligurian Apennines (16–19th C.)', in *Protoindustries et Histoire des Forêts,* ed. Jean-Paul Métaillé (Toulouse, 1990), pp. 135–50.

13 Cécile Robin and Ursula Heiniger, 'Chestnut Blight in Europe: Diversity of *Cryphonectria Parasitica*, Hypovirulence and Biocontrol', *Journal of Forest Snow and Landscape Research*, 76 (2001) pp. 361–7.

14 Carlo Montanari et al., *Note Illustrative Della Carta Della Vegetazione dell'Alta Val di Vara, Supplemento agli atti dell'Istituto Botanico e Laboratorio Crittogamico dell'Universita di Pavia, Serie 7,* 6 (1987).

15 Charles Watkins, 'The Management History and Conservation of Terraces in the Val di Vara, Liguria', in *Ligurian Landscapes: Studies in Archaeology, Geography and History*, ed. Ross Balzaretti, Mark Pearce and Charles Watkins (London, 2004), pp. 141–54.

16 Diego Moreno, 'Liguria', in *Paesaggi Rurali Storici per un Catalogo Nazionale*, ed. Mauro Agnoletti (Rome, 2010), pp. 180–203, 185.

17 See www.wildenerdale.co.uk and www.rewildingeurope.com.

Afterword

1 Michael Williams, *Americans and Their Forests: A Historical Geography* (Cambridge, 1989); George F. Peterken, *Natural Woodland* (London, 1996); George F. Peterken, *Meadows* (Gillingham, 2013).

2 Colin Tudge, *The Secret Life of Trees* (London, 2005).

3 William Cobbett, *The Woodlands* (London, 1825) paragraph 323.

4 Forestry Commission Scotland, 'Historic Woodland Survey at South Loch Katrine', www.forestry.gov.uk/histenvpolicy. The research was undertaken by Coralie Mills, Peter Quelch and Mairi Stewart.

5 Louis Gil, Pablo Fuentes-Utrilla, Álvara Soto, M. Teresa Cervera and Carmen Collada, 'English Elm is a 2,000-year-old Roman Clone', *Nature*, 431 (28 October 2004), p. 1053.

6 Robert Lowe, *General View of the Agriculture of the County of Nottingham* (London, 1798).

7 A complete history of spread of this disease and its identification, with a distribution map, is at www.forestry.gov.uk/chalara.

Select Bibliography

Baigent, Elizabeth, 'A "Splendid Pleasure Ground [for] the Elevation and Refinement of the People of London": Geographical Aspects of the History of Epping Forest, 1860–95', in *English Geographies, 1600–1950*, ed. Elizabeth Baigent and Robert Mayhew (Oxford, 2009)

Balzaretti, Ross, *Dark Age Liguria* (London, 2013)

——, et al., *Ligurian Landscapes: Studies in Archaeology, Geography and History* (London, 2004)

Bean, William J., *Trees and Shrubs Hardy in the British Isles*, 7th edn (London, 1951)

Birrell, Jean, 'Records of Feckenham Forest, Worcestershire, *c.* 1236–1377', *Worcestershire Historical Society New Series*, 21 (2006)

Cevasco, Roberta, *Memoria Verde: Nuovi Spazi per la Geografia* (Reggio Emilia, 2007)

Chambers, Douglas, *The Planters of the English Landscape Garden: Botany, Trees and the Georgics* (New Haven, CT, and London, 1993)

Coles, Bryony, and John Coles, *Sweet Track to Glastonbury: The Somerset Levels in Prehistory* (London, 1986)

Collins, E.J.T., 'Woodlands and Woodland Industries in Great Britain during and after the Charcoal Iron Era', in *Protoindustries et Histoire des Forêts*, ed. J.-P. Métaillé (Toulouse, 1992), pp. 109–20

Daniels, Stephen, and Charles Watkins, 'Picturesque Landscaping and Estate Management: Uvedale Price at Foxley, 1770–1829', *Rural History*, 2 (1991), pp. 141–70

Edlin, Herbert L., *Woodland Crafts in Britain* (London, 1949)

Elliott, Paul A., et al., *The British Arboretum: Trees, Science and Culture in the Nineteenth Century* (London, 2011)

Elwes, H. J., and A. Henry, *The Trees of Great Britain and Ireland*, (Edinburgh, 1906–13)

Evelyn, John, *Sylva* (London, 1664; 1670; 1706)

Gilpin, William, *Remarks on Forest Scenery and other Woodland Views* (London, 1791)

Griffin, Carl, 'Protest Practice and (Tree) Cultures of Conflict: Understanding
the Spaces of "Tree Maiming" in Eighteenth- and Early Nineteenth-century
England', *Transactions of the Institute of British Geographers*, 33 (2008),
pp. 91–108

Hadfield, Miles, *Landscape with Trees* (London, 1967)

Hæggström, C.-A., 'Pollard Meadows: Multiple Use of Human-made Nature',
in *The Ecological History of European Forests*, ed. Keith Kirby and Charles
Watkins (Wallingford, 1998), pp. 33–42

Hartley, Beryl, 'Exploring and Communicating Knowledge of Trees in the Early
Royal Society', *Notes and Records of the Royal Society*, 64 (2010),
pp. 229–50

Hooke, Della, *Trees in Anglo-Saxon England: Literature, Lore and Landscape*
(Woodbridge, 2010)

Jarvis, P. J., 'Plant Introductions to England and Their Role in Horticultural
and Silvicultural Innovation, 1500–1900', in *Change in the Countryside:
Essays on Rural England, 1500–1900*, ed. H.S.A. Fox and R. A. Butlin
(London, 1979), pp. 145–64

Keyser, Richard, 'The Transformation of Traditional Woodland Management:
Commercial Sylviculture in Medieval Champagne', *French Historical
Studies*, 32 (2009), pp. 353–84

Kirby, Keith, and Charles Watkins, *The Ecological History of European Forests*
(Wallingford, 1998)

Langton, John, and Graham Jones, *Forests and Chases of England and Wales,
c. 1000–c. 1500* (Oxford, 2010)

Linnard, William, *Welsh Woods and Forests* (Llandysul, 2000)

Loudon, John Claudius, *Arboretum et Fruticetum Britannicum* (London, 1838)

Meiggs, Russell, *Trees and Timber in the Ancient Mediterranean World*
(Oxford, 1982)

Métaillé, J. P., *Protoindustries et Histoire des Forêts* (Toulouse, 1992)

Mileson, Stephen, *Parks in Medieval England* (Oxford, 2009)

Moreno, Diego, *Dal Documento: Storia e Archeologia dei Sistemi
Agro-Sylvo-Pastorali* (Bologna, 1990)

——, 'Liguria', in *Paesaggi Rurali Storici per un Catalogo Nazionale* ed. Mauro
Agnoletti (Rome, 2010), pp. 180–213

Peterken, George, *Woodland Conservation and Management* (London, 1981)

——, *Natural Woodland* (Cambridge, 1996)

Petit, Sandrine, and Charles Watkins, 'Pollarding Trees: Changing Attitudes
to a Traditional Land Management Practice in Britain, 1600–1900',
Rural History, 14 (2003), pp. 157–76

Powell, J. M., '"Dominion over Palm and Pine": The British Empire Forestry
Conferences, 1920–1947', *Journal of Historical Geography*, 33 (2007),
pp. 852–77

Price, Uvedale, *Essays on the Picturesque as Compared with the Sublime and the
Beautiful: And, on the Use of Studying Pictures for the Purpose of Improving
Real Landscape* (London, 1810)

Rackham, Oliver, *Ancient Woodland* (London, 1980)

——, *History of the Countryside* (London, 1986)

——, *Woodlands* (London, 2006)

Radkau, Joachim, 'Wood and Forest in German History: In Quest of an
    Environmental Approach', *Environment and History*, 2 (1996), pp. 63–76
Ryle, George, *Forest Service: The First Forty-five Years of the Forestry Commission
    of Great Britain* (Newton Abbot, 1969)
Salbitano, Fabio, *Human Influence on Forest Ecosystems Development in Europe*
    (Bologna, 1988)
Saratsi, Eirini, et al., *Woodland Cultures in Time and Space* (Athens, 2009)
Saunders, Corinne, *The Forest of Medieval Romance: Avernus, Boceliande, Arden*
    (Woodbridge, 1993)
Schlich, W. A., *Manual of Forestry* (London, 1889–96)
Shakesheff, Tim, 'Wood and Crop Theft in Rural Herefordshire, 1800–60',
    Rural History, 13 (2002), pp. 1–17
Slive, Seymour, *Jacob van Ruisdael: A Complete Catalogue of His Paintings,
    Drawings and Etchings* (New Haven, CT, and London, 2001)
Sloman, Susan, *Gainsborough in Bath* (New Haven, CT, and London, 2002)
Smout, T. C., et al., *A History of the Native Woodlands of Scotland, 1500–1920*
    (Edinburgh, 2005)
Symonds, H. H., *Afforestation in the Lake District: A Reply to the Forestry
    Commission's White Paper of 26th August 1936* (London, 1936)
Tachibana, Setsu, and Charles Watkins, 'Botanical Transculturation: Japanese
    and British Knowledge and Understanding of *Aucuba japonica* and *Larix
    leptolepis*, 1700–1920', *Environment and History*, 16 (2010), pp. 43–71
Taylor, Maisie, 'Big Trees and Monumental Timbers' in *Flag Fen: Peterborough
    Excavation and Research, 1995–2007*, ed. Francis Pryor and Michael
    Bamforth (Peterborough, 2010)
Tittensor, Ruth, *From Peat Bog to Conifer Forest* (Chichester, 2009)
Totman, Conrad, *The Green Archipelago: Forestry in Pre-industrial Japan*
    (Berkeley, CA, 1989)
Tubbs, Colin, *The New Forest* (London, 1986)
Vera, Frans, *Grazing Ecology and Forest History* (Wallingford, 2000)
Warde, Paul, *Ecology, Economy and State Formation in Early Modern Germany*
    (Cambridge, 2006)
Watkins, Charles, *Woodland Management and Conservation* (Newton Abbot, 1990)
——, *European Woods and Forests: Studies in Cultural History* (Wallingford, 1998)
——, and Ben Cowell, *Uvedale Price, 1747–1829: Decoding the Picturesque*
    (Woodbridge, 2012)
Whyte, Nicola, 'An Archaeology of Natural Places: Trees in the Early Modern
    Landscape', *Huntington Library Quarterly*, 76 (2013), pp. 499–517
Wickham, Chris, 'European Forests in the Early Middle Ages: Landscape and
    Land Clearance', in *Land and Power: Studies in Italian and European Social
    History, 400–1200*, ed. Chris Wickham (London, 1994), pp. 155–99
Wiltshire, Mary et al., *Duffield Frith: History and Evolution of a Medieval
    Derbyshire Forest* (Ashbourne, 2005)
Young, Charles, *The Royal Forests of Medieval England* (Philadelphia, PA, 1979)

Acknowledgements

I wish to thank all members of my family and friends who have encouraged my research for this book, which is dedicated to the memory of my parents Ken and Ruth Watkins. Many friends and colleagues have provided help, assistance and advice over the years and special thanks are due to: Professor Mauro Agnoletti, Dr Pantelis Arvanitis, Dr Sallie Bailey, Dr Kostas Baginetas, Dr Ross Balzaretti, Dr Mark Bradley, Dr Clive Brasier, Dr Raffaella Bruzzone, Dr Roberta Cevasco, Dr Fiona Cooper, Professor Carl-Adam Hæggström, Sir Andrew and Lady Buchanan, Dr Harry Cocks, Dr Fiona Cooper, Dr Ben Cowell, Professor Stephen Daniels, Major David and Lindy Davenport, James Davenport, Dr Catharine Delano-Smith, Professor Paul Elliott, Professor Georgina Endfield, Dr Robert Fish, Dr Somnath Ghosal, Harry Gilonis, Dr Carl Griffin, Dr Richard Hamblyn, Dr Beryl Hartley, Dr Robert Hearn, Professor Mike Heffernan, Dr Della Hooke, Robert Howard, Dr Nuala Johnson, Dr Matthew Kempson, Dr Keith Kirby, Don Sandro Lagomarsini, Dr Jack Langton, Dr Chris Lavers, Michael Leaman, Dr Stephen Legg, Norman Lewis MBE, Dr Haydn Lorimer, Professor David Matless, Dr Peter Merriman, Professor Jean-Paul Métaillé, Dr Paul Merchant, Dr Briony McDonagh, Professor Diego Moreno, Professor Mark Pearce, Dr George Peterken, Dr Sandrine Petit, Dr Pietro Piana, Professor Pietro Piussi, Dr Clive Potter, Professor Oliver Rackham, Dr George Revill, Dr Mark Riley, Dr Graham Riminton, Jeremy Rison, Dr Eirini Saratsi, Dr Susanne Seymour, Professor Brian Short, Dr Emily Sloan, Susan Sloman, Professor Chris Smout, Jonathan Spencer, Professor Setsu Tachibana, Dr Judith Tsouvalis, Dr Alex Vasudevan, Dr Lucy Veale, Professor Frans Vera, Dr Paul Warde, Elaine Watts, Dr Philip Wheeler, David Whitehead, Professor Tom Williamson and Guy Woodford.

The following archives, libraries and galleries have been of great assistance: Abbazia di S Colombano, Bobbio; Arni Magnusson Institute, Reykjavík; the Athenaeum Library, London; Bayerische Staatsgemäldesammlungen, Neue Pinakothek, Munich; Bodleian Library, Oxford; British Library, London; British Museum, London; Musée Condé, Chantilly; Courtauld Institute of Art Gallery, London; Forest History Society, Durham, North Carolina; Hallward Library, University of Nottingham; Herefordshire Record Office; Linnean Society of London; Fondation Custodia, Frits Lugt Collection, Paris; Kunsthalle Hamburg; National Gallery of Art, Washington, DC; Norfolk County Council; Paul Mellon Centre for

Studies in British Art, London; Public Record Office of Northern Ireland, Belfast; Metropolitan Museum of Art, New York; National Gallery, London; National Portrait Gallery, London; Soane Museum, London; Society of Antiquaries of London; South Tyrol Museum of Archaeology, Bolzano; Collection of Earl Spencer, Althorp; Staatliche Museen, Berlin; Thoresby Estates, Nottinghamshire; Tate, London; Whitworth Art Gallery, Manchester; National Art Library, Victoria & Albert Museum, London.

I would also like to thank all my colleagues in the School of Geography, University of Nottingham, for their help and support and the undergraduates and research students at Nottingham who provide such a rich fund of knowledge and enthusiasm. Of course, the best way of learning about trees, woods and forests is to study and work with and in them. Hundreds of woods in Britain owned by organizations such as the National Trust, the County Wildlife Trusts and the Woodland Trust are open to the public. Many insights into woodland management and history can be gained by joining the Royal Forestry Society and attending the woodland visits they organize every year.

Photo Acknowledgements

The author and publishers wish to express their thanks to the below sources of illustrative material and/or permission to reproduce it:

Althorp House, Northamptonshire (Spencer Collection): 46; Árni Magnússon Institute for Icelandic Studies, Reykjavík (Ms. AM 738 4to): 16; from Aaron Arrowsmith, *Plan of the Arboretum in the Garden of the Horticultural Society at Chiswick, March 1826* (London, 1826): 31; photos author: 3, 51, 52, 53, 54, 65, 66, 77, 80, 99, 102, 105, 107, 111; photos Ross Balzaretti: 106, 109; British Museum, London (photos © Trustees of the British Museum): 1, 4, 7, 8, 9, 10, 11, 12, 13, 14, 15, 18, 20, 21, 36, 37, 38, 39, 40, 43, 47, 48, 49, 50, 55, 67, 68, 69, 72, 73, 75, 81; from E. Adveno Brooke, *The Gardens of England* (London, 1857): 86; from James Brown, *The Forester: A Practical Treatise on the Planting, Rearing, and General Management of Forest Trees…*(Edinburgh and London, 1871): 85; from Mrs Comyns Carr [Alice Vansittart Strettel Carr], *North Italian Folk: Sketches of Town and Country Life* (London, 1878): 110; Derby Library (photos © Derby Museums Trust): 58, 84; Fondation Custodia, Frits Lugt Collection, Paris: 41; reproduced courtesy of the Forest History Society, Durham, North Carolina: 91, 92, 98; from [Robert Goadby], *A New Display of the Beauties of England: or, a Description of the most Elegant or Magnificent Public Edifices, Royal Palaces, Noblemen's and Gentlemen's Seats, and other Curiosities, Natural or Artificial, in different Parts of the Kingdom . . .* (London, 1773): 28; from Horace G. Hutchinson, *Shooting* (London, 1903): 88, 89, 90; Imperial War Museum, London (photo © Imperial War Museum): 78; from A. H. Kent, *Veitch's Manual of the Coniferæ: containing a general review of the order; a synopsis of the species cultivated in Great Britain; their botanical history, economic properties, place and use in arboriculture, etc., etc.* (London, 1900): 87; Kunsthalle Hamburg: 42; Los Angeles County Museum, reproduced by permission of Los Angeles County Museum of Art (LACMA) Image Library (image Museum Associates/ LACMA): 97; from John Claudius Loudon, *Arboretum et Fruticetum Britannicum: or, the Trees and Shrubs of Britain, Native and Foreign, Hardy and Half-hardy, Pictorially and Botanically Delineated, and Scientifically and Popularly Described . . .* (London, 1838): 24, 25, 26; from John Claudius Loudon, *An Encyclopædia of Gardening: comprising the theory and practice of horticulture, floriculture, arboriculture, and landscape-gardening, including all the latest improvements . . .* (London, 1830): 83; from Robert Monteath, *The Forester's*

 Index

Abbotsford 79, 141
Ablett, William 181
Acer campestre 123, 182, 274
Acer criticum 124
Acer negundo 67
Acer platanoides 18
Achaeans 31
Acland, Francis Dyke 221–2
acorns 26–8, 32, 54, 94, 119, 126–7,
    206, 248, 256
Adam of Bremen 48
Aelfric 50
Aeneas 40–42
Ætheling, Edgar 52
Agamemnon 31
agricultural depression 199, 219
agricultural intensification 269
Aias 31
Ajax 31
Alcinous 32
Alcock, Rutherford 85
alder 19–23, 119, 121, 128, 131, 177,
    223, 252, 269, 271, 274
Alexander the Great 13, 29, 36
Alfonso XII, King of Spain 201
Altdorfer, Albrecht, *Landscape with
    Spruce and Two Willows* 95, 37
Ancient Tree Forum 139
ancient woodland 12, 21, 159–60,
    172, 180, 203
Anderson, James 133
Anne, Queen 65
Antilochus 32

Antinous 38
Apollinaris, Domitius 25
arboreta 73–7
Arcadia 29
Argyll, 3rd Duke of 72
Aristotle 29
Arnold Arboretum 83, 213, 231
Artaxerxes I, King of Persia 43
ash 15, 18, 21, 29–31, 43, 46–8, 70,
    90, 115, 119, 121, 128–31, 135,
    182, 184, 191, 204, 252, 254,
    257–8, 265, 271–4
Asius 30–31
assarting 57, 59
Athanasius, St 50
Atholl, Dukes of 175, 184
Atkinson, William 79
Atlas, Mount 78
'Auraucaria Avenue at Bicton, The' 87
Austen, Jane, *Mansfield Park* 176–7, 198

Bagot, Ralph 60
Banister, John 66
Bankes, Richard 143
bark 17–18, 24, 26, 30, 85, 89–91,
    94–5, 98, 113, 150, 157, 182,
    233, 274
Barron, William 193
Bartram, John 73
Bathurst, Lord 72
Bean, William Jackson 76
bear 37–8, 50, 122
Bedford, 6th Duke of 76, 183

beech 18, 28, 31, 43, 70, 94, 100, 104,
    107, 113, 121, 128, 131, 176,
    184–6, 206–7, 223
beetles 138, 159, 173, 274
Bellini, Giovanni 95
    Assassination of St Peter the Martyr
    76
    Madonna of the Meadows 71
Bentham, Jeremy 179
Bewick, Thomas 242
    Red Deer Stag 68
Biddulph, John 135
Bierstadt, Albert, *Grizzly Giant Sequoia*
    97
Big Basin Redwoods Park 225
Bilhagh *144*, 145–8, 153, 160
Biltmore 214
birch 18–19, 140, 145, 159, 171, 271,
    274
Birklands *144, 145,* 145–8, 153, 155–6,
    160, *172*
Birnam Wood *167*, 175
blackthorn 128, 197, 259, 260
Blake, Thomas 242
Bledisloe, Lord 234
boar, wild 15, 32–3, 37–9, 55–6, 267,
    20, 21
Board of Agriculture 115, 132, 217–18
Bohun, Humphrey de 62
Borysthenes 38
box 33, 157, 193
Boy with Horse 7
bracken 153, 159, 173, 222
Brandis, Sir Dietrich 210–14, *211*
breck 140, 145
Breckland 11, 222
Breuteuil, Evrard de 50
Bright, Tom 181
British Mycological Society 159
Brown, James 190–92, *192*
Brown, Lancelot 'Capability' 108
browsing 9, 120–22, 205, 250
Bruegel, Pieter, *Woodland Scene with*
    Bears 96, 38
Brydone, Patrick 91–3
Bucephalus 37
Bunbury, Henry William, *Hop Pickers*
    81

Burke, Edmund 107
Burnham Beeches 118
Byron, Lord 141, 152

Caldecott, Randolph, *Gathering the*
    Chestnuts 110
Call, John 144
Callitrus quadrivalus 33
Canadian Forestry Corps 221
Cantalupo, Margery de 60
carbon sequestration 268
Carlos I, King of Portugal 201
Carlowitz, Hanss Carl von 204
Carr, Alice Comyns 262
Carr, John 157
Carracci, Annibale, *Garden of Eden 13*
Carter, James 155
Castagno de Cento Cavalli 91–2
Catesby, Mark 68
Cato, Marcus Porcius 28
cattle 28–9, 32, 73, 120, 129, 133,
    173, 205, 252–4, 269–71
cedar 30, 36, 72–8, 194, 202–3
Central Park 229
Chalara fraxinea 90, 274
Chambers, Thomas, *Mount Auburn*
    Arboretum 33
Champagne, Countess Blanche of 207
charcoal 19, 50, 121, 178, 180, 262
Chatsworth 80
Cheiron 31
chestnut, sweet 15, 29, 70, 92, 128,
    131, 171, 182, 247, 254–66
Christ 48, 99
Cicero 34
Circe 32
Clark, Galen 229
Clark, John 132
Clarke, George 145–7
Claude 105–9, 114, 187
clearance 7, 15, 59, 137, 169
Clipstone Archway School 61
Clumber Park 148
clumps 8, 77–80, 179, 187, 197–9
coal 133, 136, 160, 180, 221
Cobbett, William 270
Coleman, William 77
Collinson, Peter 73

Columbanus, St 248
Columella, Lucius Junius 29, 271
common land 119, 128, 137, 205,
    264, 269
Compton, Henry 65–8
Conness, John 228
Constable, John 102, 105
Cook, Moses 129, 131
coppice 26–8, 114, 177, 180–82, 190,
    200–209, 216–17, 264–5, 271–3,
    112
coppicing 11–13, 26, 181–2, 206–8,
    264, 269, 273
Corydon 43
Costa dei Ghiffi *102*
Cotta, Heinrich 209
Coughton, Constance of 60
Council for the Preservation of Rural
    England 237, 241
Courthope, Sir George 235–7
Coventry, Andrew 179
Cowper, William 177
Cozens, John Robert 92
Craig, William 117
Cryphonectria parasitica 263
Cryptomeria japonica 195
Cullum, Rev Sir Thomas Gery 117
cultural landscape 15, 268
Cupressus macrocarpa 77
Cytisus scoparius 265

Dalhousie, 10th Earl of 210
dead wood 173, 179
Death of Euryalus 11
Death of William Rufus 18
Dedication Scheme 169
deer 18, 37, 42, 52–62, 115, 129,
    146–8, 156, 205, 209, 267
Dehra Dun 212
dendrochronology 8, 22, 40, 150, 173,
    224, 256, 260, 271
Derby Arboretum 80–82
Despenser, Almeric le 60
Diana 39–40
Domesday Book 177
Douglas, David 242
Douglas fir 87, 89, 195, 233, 241, *24*
Downing, Andrew Jackson 82–3, 231

Downton 92
druids 149
Dryden, John 69
Dukes Wood 271–5, *272*
Duleep Singh, Maharajah 201
Dunkeld 88, 184
Dunstall, John, *Pollard Oak 72*
Dürer, Albrecht 95–6
    Adam and Eve 14
    Pine Trees 36
    Spruce 69
Dutch elm disease 10, 274, 286

Eastnor 67, 77, *77*, 182, 195
Eden 43, 45, 99, 101
Edlin, Herbert 238, 240–45
Edward I, King 56, 62
Ellis, William 131
elm 15, 19, 21, 28–30, 46, 71, 80, 119,
    128–35, 184, 191, 195, 271, 274
Elvaston Castle 77, 80, 193, *194*
Elwes, Henry John 86–7
Emerson, Ralph Waldo 226–30
Ennerdale 267
Epping Forest 128, 137, *166*
Erica arborea 253, 265
Etna, Mount 91–2
EU Habitats Directive 173
Euryalus 42
Evelyn, John 130–31, 150–51, *27*
excarnation 25

faggots 94, 121, 159
Farington, Joseph 184
Father of the Forest Tree *95*
Feckenham Forest 55–63
fig 27–8, 250
firewood 28–9, 59, 119, 121, 130–37,
    147, 153, 179, 180, 204–6, 229,
    254–7, 273–4
First World War 14, 203, 215, 219
Fisher, W. R. 212
fitz Nigel, Richard 62
Flag Fen 20–24
Fomes fomentarius 17
Forbes, A. C. 190, 216
Fordyce, John 144
Forest Charter 54–5

Forest of Dean 10, 52, 56, 143, 146, 177, 216–18, 234–40
Forestry Commission 10–12, 88–9, 160, 169–73, 222–3, 234–42, 267, 271, 274
Fortune, Robert 77
fox hunting 14, 177, 196–7, 203
Foxley 103, 114
Franz Ferdinand, Archduke 201
Frazer, Sir James 40
Fuji, Mount 85–6
Fulham Palace 65, *66*

Gainsborough, Thomas 93, 102–5
    Beech Trees at Foxley 45
    William Poyntz 46
Gastineau, Henry, *Hafod House 82*
Gayer, Karl 212
Geerten tot Sint Jons, *John the Baptist 17*
Genista salzmannii 264
Gilpin, William 93, 108–15, 152, 196
    Pollard 49
    Unbalanced Tree 48
Gilpin, William Sawrey 189
Gladwin, John 145
Glasnevin 73, *29*
Glastonbury Abbey 21
goats 8, 29, 37, 43, 73, 120–28, 249–50
Goethe, Johann Wolfgang von 92
Gooch, Mrs Elizabeth Sarah Villa Real 141
Gore, Charles 92
Gray, Thomas 108
grazing 9, 11, 14, 23, 120–22, 127, 145–7, 170–73, 200–207, 222, 242, 247, 255–69
Grindal, Edmund 65
Grizzly Giant Sequoia *228*
Gwydyr Forest 241

Hackert, Jakob Philipp 92
Hadfield, Miles 240
Hadrian, Emperor 13, 36, 38–9
Hafod *185*, 190
Hall, Spencer 152
Hardwick Hall 149
Hardy, Thomas 182
Harriman, W. A. 246

Harrington, Earls of 193
Hartig, Georg Ludwig 210
Hatfield Forest 128
Havelock, General Henry 211
Havelock, Rachel 211
hawthorn 119, 128, 131, 195, 256
hay 119, 121, 253–60, 265, 269
Hayachine Mountain 9, *161*
hazel 18–22, 28, 119, 182, 260, 271
Hearne, Thomas, *Oak Trees, Downton 134, 73*
heathlands 11, 222, 269
hedgerows 10, 13–14, 52, 104, 115, 119, 128–36, 177–9, 184–5, 190, 195, 203, 271, 274
Henry I, King 54
Henry II, King 54
Henry III, King 55
Hess, Richard 212
Hill, Andrew 225
Hillingdon 70
Hitchcock, Alfred 224
Hogarth, William 107
Holford, Robert 195
Holkham shooting plan *88*
Holroyd, Charles, *Eve and the Serpent 1*
holly 21, 128, 193
Holt, George 170
Homer 30–32
hop poles 181, 272
hornbeam 28, 71, 122–3, 128, 131, 138, 252–60
hornbeam, hop 19, 123, 252, 256–61
Horticultural Society 79–80, 242
Houël, Jean-Pierre 92
Hovey, Charles 82
Howitt, William 152
Humboldt, Alexander von 226
hunting 10, 13, 26, 32–55, 60–63, 94, 143, 148, 183, 196–9, 205–8, 216–19, 230, 240, 271

Ice Man 17–19
Idomeneus 30
Indian Forest Service 212
Interstate Palisades Park 246
Irving, Washington 141, 152
ivy 33, 115, 131, 261

James II, King 65
Jerome, St 96
John I, King 54, 63
Johnes, Thomas 185
Johnson, Samuel 91, 175
Jones, Noah 81
Jones, Thomas, *Dido and Aeneas* 10
    On the Banks of Lake Nemi 67
Judeich, Friedrich 212
juniper 43, 122, 193–4, 256–7, 265

Kaempfer, Engelbert 84–5
Kalm, Pehr 72
Katrine, Loch 271
Kent, Nathaniel 136
Kent, William 107, 231
Kew 71, 76
Kielder Forest 240–42
Killerton redwoods *232*
Kin sen shou 34
Knight, Richard Payne 92, 102, 185

Labour Party 172
Lagorara Valley 108
Lake District 10, 108, 237, 240–41,
    267
land abandonment 14, 247, 256–9,
    264–7
Langley, Batty 129
larch 13, 18, 85–9, 176, 184–91, 203,
    223, 240–41, *26*
Laudine, Lady 50
Lawson, William 112
leaf fodder 13, 115, 119–28, 133–6,
    247–50, 266
Leonteus 31
Lievens, Jan, *Landscape with Three
    Trees 41*
Limbourg Brothers, *December 35*
lime 17, 21, 29, 71, 115, 119, 123
Lincoln, Abraham 229
Lindley, John 85, 227
Linnaeus, Carl 71–3
Linnard, William 190
lion 36–8, 98, 155
Liquidambar styraciflua 66
Lloyd George, David 217
Lobb, William 77, 227

Loddiges, George 79
London, George 66
London, Jack 230
lopping 121, 130–39, 147, 256
Lopping Hall *138*
Loudon, John Claudius 65, 69, 74, 80,
    179, *188*
Lucan 34
Lutz, Frank 245
Lycia 27

Mackley, George 241–2
    In the Sprucewoods 100
    Border National Forest Park 101
Magnolia virginiana 67
Main, James 117, 180
Major Oak 155, *158*, 160
Mantegna 95
Manvers, 3rd Earl 155–59
Manwood, John 55
Mariposa Grove 226
Marshall, William 131
Martin, John, *Fall of Man 12*
Mason, William 108, 110
meadows 26–8, 35, 50, 121, 268–9
Meiggs, Russell 26–7
Meliboeus 43
Menzies, Archibald 242
Middleton, Charles 144
Middleton, John 132, 136
Miller, Philip 71
Millhouse, Robert 152
Milton, John 187
mistletoe 40, 42, 149
Monteath, Robert, 'Coppice Trees' *112*
Mornington, 4th Earl of 137
Mortimer, John 130
Mortimer, Roger 63
Morton Bagot Church 22
Mount Auburn 82, *83*
Mucianus, Licinius 27
Muir, John 225–34
Muir National Monument *226, 232,*
    233

Nash, Paul, *We Are Making a New
    World 220, 78*
National Forest Park 10, 235–41

National Park 229–31, 236–7, 246
nature trails 245
Nebuchadnezzar 36
Nehemiah 43
Nemi 39–40
Nestor 32
Neville, Hugh de 63
New England Sawmill Unit 221
New Forest 10, 52–3, 109–15, 143,
    146, 177, 218, 236
Newcastle, 1st Duke of 148
Newfoundland Forestry Corps 221
Newstead Abbey 141, *55*
Nicholson, George 76
Nidhogg 46
Nisbet, John 182, 212
Nisus 42
Norway spruce 184, 223
Nottingham Arboretum *32*
Nottinghamshire Wildlife Trust 271
Novak, Kim 224, *225*
November, San Colombano, Bobbio *103*
Nuneham, 'Beaters Crossing the
    Bridge' *90*

oak 18–30, 40, 43, 59, 70, 80, 91, 99,
    101, 112–51, 156–60, 169–73,
    177, 182–6, 191, 195, 207, 233,
    248–56, 264–6, 271, 274
Òðinn 47–8
Odysseus 32
olives 26, 261
Olmsted, Frederick Law 83, 214,
    229–33
Olympic Games 33
Otho, King of Greece 210
Ottley, William, *Study of Trees and
    Rocks, Ariccia 8*

Paliama Monumental Olive Tree,
    Crete *65*
Pankrates 38
pasture 9–14, 17, 28, 35–7, 52–7, 93,
    119, 123, 143, 159, 170–75, 183,
    205, 245–50, 256–8, 268
Petre, 9th Lord 72
Peziza Willkommii 87
pheasant 176, 196–203

Philip II, King of Macedonia 36–7
Philip II, King of Macedonia 37
Philip IV, King of France 207
Phillyrea latifolia 124–6
Phytophthora ramorum 89–90
Picea abies 95
Picea sitchensis 11, 87, 223, 242–5
Picturesque 13, 102–8
pigs 28, 32, 54–9, 73, 94, 153, 206, 248
Pinchot, Gifford 213–14
pine 8, 18–19, 28, 30, 43, 71, 85–7,
    95, 121, 151, 169, 177, 184–95,
    203, 223, 229, 241
Pinus insignis 77
Pinus nigra 160, 167
Pinus pumila 9
Pinus sylvestris 95, 169, 191–3, 203
Pitt, William the Younger 116
plane 27, 32, *25*
plantations 8, 11–15, 26–7, 80, 85–7,
    114, 148, 151, 160, 169–203,
    216–17, 220, 223, 236–45, 250,
    267
Pliny the Elder 23, 26, 34
Pliny the Younger 25, 26, 33
Pococke, Richard 72
pollard 93, 96–105, 115–36, 144,
    175–7, 248, 253, 257, 262, 271
pollarding 13, 93, 115–38, 147, 250,
    253, 256, 269
Polypoetes 31
Pomeroy, W. T. 132
poplar 21, 28–32, 43, 71, 121, 131
Portland, 3rd Duke of 148, 155–7, 201
Poussin, Nicolas 109, 187
Price, Uvedale 88, 93, 103–16, 133–6,
    152, 177–89
Price, Uvedale Tomkyns 103
Pterocarpus erinaceus 121
public access 11–12, 56, 171, 178, 197,
    230–45, 258

Quercus cerris 123, 248, 255–6
Quercus coccifera 124
Quercus frainetto 123
Quercus ilex 65, 124–5, *23*
Quercus petraea 123, 252
Quercus robur 123

Ramblers' Association 11, 237
Ray, John 65–6, 110
Recknagel, Arthur 215
Recupero, Guiseppe 91–3
Rembrandt 96–9
   St Jerome 39
   Three Trees 40
Repton, Humphry 116, 149
re-wilding 247, 267
Reynardson, Samuel 70
Rhododendron ponticum 89, 203
Richard I, King 54, 140, 155
Richmond, 2nd Duke of 72
Robin Hood 140–43, 151, 155, 172
Robinia pseudoacacia 90, 110, 270, *78*
Robinson, Roy 218–221, 235–6
Rodgers, Joseph 160
Rooke, Hayman 149–55
   Plantations at Welbeck 58
   Turkish Kiosk in Plantation,
   Farnsfield 84
Rooke, Noel 241
Roosevelt, Theodore 214, 230, 233
Rosa, Salvator 105–6, 109, 112
   Mercury and the Dishonest
   Woodsman 70
   Study of Trees 47
rowan 119, 121
Royal Boys 37
Royal Forestry Society 170, 202, 217
Royal Forests 10, 52–61, 143–4, 218
Royal Indian Engineering College 212
Royal Saxon Forest Academy 210
Rubens, Peter Paul, *A Boar Hunt 20*
Rufford 148, 155
Rufus Stone 53
Rufus, Calvisius 26
Rufus, Quintus Curtius 38
Ruisdael, Jacob van 93, 100–103, 109,
   157
   Forest Marsh 44
   Landscape with a Cottage and Trees
   42
   The Three Oaks 43
Ruskin, John 7, 202–3
Russian Cabin, Sherwood 62
Ruysdael Oak, Sherwood 64
Ryle, George 219, 222

Salvin, Anthony 156
San Francisco Bohemian Club 230
Sandby, Paul, *Mr Whatman's Paper*
   Mill 50
Sanderson, George 272
Sargent, Charles 76, 86, 213, 231
Schenck, Carl Alwin 214, *215*
Schiller, Friedrich 210
Schlich, Sir William 210–18, *215*, 221
Scott, Sir Walter 79, 93, 140–43,
   151–2, 156, 271
Seahenge 25, *5*
Searle, January 153–7
Second World War 122, 160, 169,
   171, 265
semi-natural vegetation 9, 267
Sequoia giganteum 226–30
Sequoia sempervirens 9, 195, 224–32,
   225, 226, 227
sheep 8, 17, 28–9, 37, 73, 120–33,
   147, 173, 205, 221, 249–50, 258,
   265–71
shepherds 124–28, 249
Sherwood Forest 11, 14, 118, 140–60,
   169–74, 222, *79, 80*
Shield, William 141
shooting 14, 113, 176–7, 196–203,
   216–19, 267
shredding 115, 119–37, 247–55, 267
Siebold, Franz von 84–6
Sierra Club 229, 233
Simoeisius 31
Simpson, John 199
Sinclair, George 76, 183
Slow Food Association 266
Snowdonia National Forest Park 237
Somers, 2nd Earl 67, 78, 195
Somerset Levels 20–22
Stanage Game Plan 89
Standish, Arthur 130
Steuart, Sir Henry 117
Stewart, James 224, *225*
Stoddart, John 175
Strabo 34
Strutt, Jacob 117
Strutt, Joseph 81–2
Stubbs, George, *Freeman, the Earl of*
   Clarendon's Gamekeeper 19

Sturluson, Snorri 45–6
Sweet, Raymond 21
Sweet Track 20–21, *4*
Switzer, Stephen 65, 70

Tamalpais, Mount 230–31
tax 8, 56, 169, 207
terraces 14, 247–68
Theophrastus 23, 29–30
Thomsen, Peter 234
Thomson, Christopher 153
    'The Major Oak' *59*
    '"Simon the Forester" *60*
Thoreau, Henry David 226
Thoresby 148, 156, 160, 169–73
Thunberg, Carl Peter 84–5
Thyrsis 43
timber 12–14, 23–34, 43, 55–9, 68,
    87–90, 109–36, 140, 144–6, 159,
    169, 177–9, 183, 186, 191–3,
    200–223, 231–46, 253, 267, 274
Tolstoy, Leo 198
Trajan 26, 38
transhumance 123, 249–50
Trojans 31–2, 42
Troyes, Chrétien de 50
Trueman, John 157
Tsuga diversifolia 9
Turgenev, Ivan 156, 198
Turner, Joseph Mallord William,
    The Golden Bough 9

Ull 48
Ulmus glabra 10, 271, 274
Ulmus procera 10, 271, 274

Valletti, willow pollards at *109*
Vanderbilt, George W. 214
Varese Ligure *251–2, 257*
Varnier, Hans the Elder, *Tree of
    Knowledge 15*
Varro, Marcus Terentius 29
Vaux, Calvert 231
Veitch, John Gould 77, 85–7, 195,
    227
Vergina 36, *6*
vert 56
Vertigo (film) 224, *225*

Viburnum lantana 18
Viburnum tinus 88
Victoria, Queen 137
vines 26–9, 250, 255, 260
Virgil 43, 69

Wade, Walter 73
Walpole, Horace 107–8
Wardian case 67
Washingtonia 227
Watkins, Carleton, *Grizzly Giant 96*
Waugh, Evelyn 83
Welbeck 148, 155, *157–8*, 201
Wellington, 1st Duke of 137, 227
Wellingtonia 227
West, J. 180
Westonbirt 77, 195
Whellens, W. H. 159
Whitaker, Joseph 159
Whitton Park *28*
wilderness 48, 176, 236–8, 241
Wilhelm II, Kaiser 201
William, Duke of Normandy 52
William of Malmesbury 52
William Rufus 52
Williams, Hugh William 'Grecian',
    Birnam Wood 75
Willingale, Samuel and Alfred 137
willow 19–21, 28, 93–105, 117, 121,
    128–31, 250, 260, 269
wolf 27, 50, 122
Women's Forestry Corps 221
wood pastures 56, 123, 205, 255, 271
Wordsworth, William 93, 152, 226,
    233, 240
Worksop Manor 72, 148
Worlidge, John 129–30
    *Wounded Giant, Sherwood Forest
    170*
Württemberg, Duke of 208

yew 18, 48, 193–4
Yggdrasill 46–9, *16*
Yosemite 226–9, 234–5, 246
Young, Arthur 115, 133
Yvain 50

Zaros, Crete *125, 127*